T0135919

Proceedings

of the

5ᵗʰ International Beilstein Symposium

on

EXPERIMENTAL STANDARD CONDITIONS OF ENZYME CHARACTERIZATIONS

Protein Structure Meets Enzyme Kinetics

September 12ᵗʰ – 16ᵗʰ, 2011

Rüdesheim/Rhein, Germany

Edited by Martin G. Hicks and Carsten Kettner

Beilstein-Institut

Experimental Standard Conditions of Enzyme Characterization
September 12th – 16th, 2011, Rüdesheim/Rhein, Germany

BEILSTEIN-INSTITUT ZUR FÖRDERUNG DER CHEMISCHEN WISSENSCHAFTEN

Trakehner Str. 7 – 9
60487 Frankfurt
Germany

Telephone:	+49 (0)69 7167 3211	**E-Mail:**	info@beilstein-institut.de
Fax:	+49 (0)69 7167 3219	**Web-Page:**	www.beilstein-institut.de

IMPRESSUM

Experimental Standard Conditions of Enzyme Characterizations – Protein Structure Meets Enzyme Kinetics, Martin G. Hicks and Carsten Kettner (Eds.), Proceedings of the 5th Beilstein-Institut Symposium, September 12th – 16th 2011, Rüdesheim, Germany.

Bibliographic information published by the *Deutsche Nationalbibliothek*. The *Deutsche Nationalbibliothek* lists this publication in the *Deutsche Nationalbibliografie*; detailed bibliographic data are available in the Internet at http://dnb.ddb.de.

ISBN 978-3-8325-3367-0

Layout by: Hübner Electronic Publishing GmbH
Steinheimer Straße 22a
65343 Eltville
www.huebner-ep.de

Printed by: Logos Verlag Berlin GmbH
Comeniushof, Gubener Str. 47
10243 Berlin
www.logos-verlag.de

Cover Illustration by: Bosse und Meinhard GbR.
Kaiserstraße 34
53113 Bonn
www.bosse-meinhard.de

 Beilstein-Institut

III

Experimental Standard Conditions of Enzyme Characterization
September 12th – 16th, 2011, Rüdesheim/Rhein, Germany

PREFACE

The multi-disciplinary approaches which are now typical of research in the natural and life sciences use a combination of modern experimental techniques which have led to an increased accuracy in the measurements of enzyme structures and activities. Modern analysis methods often result in the generation of huge amounts of data and large data sets, which are subsequently published in electronic data repositories and in publications.

However, data in both the literature and in databases suffer from the fact that they are often non-comparable due to incomplete and imprecise descriptions of materials and methods. Furthermore, if the experimental conditions are not fully and accurately stated, the values of the functional data of enzyme activities will be of little use for applications such as systems biology.

Further problems occur even when the data are well reported; they will have often been collected under quite disparate conditions so that researchers are faced with the problem of the range of method-specific enzyme data. This is often an issue when data move between researchers whose data are supplied by laboratories that use different methods, and can, in the worst case, lead to misinterpretation of laboratory findings.

Since 2003 the STRENDA Commission (Standards for Reporting Enzyme Data) has been actively working on concepts to improve the quality of reporting functional enzyme data that will allow the efficient use of enzyme kinetics in the *in vivo*, *in vitro* and *in silico* investigation of biological systems. The Commission has two major goals: the first is the development of a set of guidelines for the reporting of data in publications. These guidelines are currently recommended by 28 biochemistry journals. The second goal is the development of an electronic data submission tool that incorporates the STRENDA Guidelines, and which is intended to act as a portal for the submission of enzyme kinetics data to a freely-accessible, public database.

The previous four ESCEC symposia have not only supported the work of the STRENDA Commission, but have also lead to the symposium becoming established as a scientific meeting in its own right. For the fifth symposium, the organizers decided to choose 'Protein Structure Meets Enzyme Kinetics' as the focus of the meeting – providing a perfect example of an interesting and important area of contempory science where the reporting of data needs to be improved. The characterization of enzyme functions is usually accompanied by the determination of the rates by which enzymes catalyze reactions. This knowledge can give insights into the mechanism of the reaction, which in turn relates to the structure of the enzyme.

At the 5th ESCEC Symposium, organized by the Beilstein-Institut together with the STRENDA Commission, we were fortunate that a diverse range of speakers not only accepted our invitation to present but that most were able attend for the whole meeting and so participate in the lively discussions. Topics covered ranged from describing how the modification of enzyme structures affects the kinetics of enzyme reactions to discussing new results, approaches and methodologies for establishing physiological ties between sequence, structure and kinetics and modeled networks of collaborative enzymes. The overall topic – standard representation of enzyme data – was considered and discussed in precise detail when the initial version of the STRENDA capturing

tool was presented. This has enabled the Commission to complete a working data acquisition prototype, which can be accessed at http://www.strenda.org/eform.html/. Any comments and suggestions are still welcome!

We would like to thank particularly the authors who provided us with written versions of the papers that they presented. Special thanks go to all those involved with the preparation and organization of the symposium, to the chairmen who piloted us successfully through the sessions, and to the speakers and participants for their contribution in making this symposium a valuable and fruitful event.

Frankfurt/Main, February 2013

Carsten Kettner
Martin G. Hicks

Beilstein-Institut

Experimental Standard Conditions of Enzyme Characterization
September 12th – 16th, 2011, Rüdesheim/Rhein, Germany

V

CONTENTS

Phosphonates to Phosphate: A Functional Annotation of the Essential Genes of the *Phn* Operon in *Escherichia coli*

Siddhesh S. Kamat, Howard J. Williams, and Frank M. Raushel[*]

Department of Chemistry, P.O. Box 30012, Texas A&M University,
College Station, TX 77843, U.S.A.

E-Mail: *raushel@tamu.edu

Received: 16th February 2012/Published: 15th February 2013

Abstract

The reaction mechanism for the enzymatic conversion of methyl phosphonate to phosphate and methane in *Escherichia coli* has eluded researchers over the last three decades despite significant genetic and *in vivo* studies. The *phn* operon governs the C-P lyase activity in *E. coli*. The essential genes within the *phn* operon are *phnGHIJKLM*. The proteins encoded by *phnGHM* were over-expressed in *E. coli* and purified to homogeneity using standard protocols. The proteins encoded by *phnIJKL* were soluble only when expressed as N-terminal glutathione S-transferase (GST) fusion proteins. PhnI was shown to catalyse the formation of α-D-ribose-1-methylphosphonate-5-triphosphate (RPnTP) from MgATP and methylphosphonate in the presence of PhnG, PhnH, and PhnL after *in situ* cleavage of the GST-tags. PhnI alone catalyses the hydrolytic cleavage of MgATP to adenine and D-ribose-5-triphosphate (RTP). PhnM catalyses the hydrolysis of α-D-ribose-1-methylphosphonate-5-triphosphate (RPnTP) to α-D-ribose-1-methylphosphonate-5-phosphate (PRPn) and pyrophosphate with attack of water on the α-phosphoryl group of the triphosphate moiety of RPnTP. PhnJ was reconstituted with an iron-sulphur cluster through the anaerobic addition of $FeSO_4$, Na_2S and $Na_2S_2O_4$ under strictly anaerobic conditions. The $[Fe_4S_4]$-reconstituted PhnJ

GST-fusion protein catalyses the homolytic cleavage of the phosphorus-carbon bond of α-D-ribose-1-methylphosphonate-5-phosphate (PRPn) to ultimately form α-D-ribose-1,2-cyclic-phosphate-5-phosphate (PRcP) and methane in the presence of *S*-adenosyl-L-methionine (SAM) under strictly anaerobic conditions, after *in situ* cleavage of the GST-tag.

INTRODUCTION

Carbon, nitrogen, oxygen, and phosphorus are essential elements required by all living organisms for survival. Of these elements, phosphorus is primarily incorporated as a derivative of phosphate where it is an integral component of nucleic acids, carbohydrates, and phospholipids. Phosphorus has many roles in many metabolic pathways and cell signalling. Most organisms have evolved to obtain phosphorus, as phosphate, directly from the environment. However, many Gram-negative bacteria, such as *Escherichia coli*, have the capability of utilizing organophosphonates as a nutritional source of phosphorus under conditions of phosphate starvation via the cryptic C-P lyase activity (1, 2). Phosphonates are organophosphorus compounds that contain one C-P bond. This bond is chemically inert and hydrolytically stable. Phosphonates are ubiquitous and some prominent examples include antibiotics (fosfomycin and phosphonothrixin), herbicides (glyphosate) and industrial detergent additives (amino-trimethylene phosphonate). It is estimated that more than 20,000 tons of phosphonates are released annually in the US alone from herbicide and detergent wastes (3, 4). The large quantities of phosphonates that are being released into the environment suggest that there needs to be a greater understanding of how these compounds are metabolized and degraded.

In *E. coli* the catalytic machinery for the "C-P lyase" activity has been localized to the *phn* gene cluster, which is induced only under conditions of limiting phosphate by the global *pho* regulation system. This operon consists of 14 genes denoted as *phnCDEFGHIJKLMNOP*. Genetic and biochemical experiments have demonstrated that the *phnGHIJKLM* genes encode proteins that are essential for the enzymatic conversion of phosphonates to phosphate. The genes *phnCDEF* encode proteins responsible for the binding and/or transport of phosphonates across cell membranes as well as regulation and modulation of the C-P lyase activity. The genes *phnNOP* encode proteins that have auxiliary functions and these three proteins have been functionally characterized (1, 5). Wanner, Hove-Jensen and co-workers have shown that PhnN catalyses the ATP-dependent phosphorylation of α-D-ribose-1,5-bisphosphate (PRP) to α-D-ribose-1-pyrophosphate-5-phosphate (PRPP) (6). Blanchard and co-workers have shown that PhnO catalyses the metal-dependent acetylation of aminoalkylphosphonic acids with acetyl CoA as the co-substrate (7). Zechel, Hove-Jensen and co-workers have shown that PhnP catalyses the metal-dependent hydrolysis of α-D-ribose-1,2-cyclic phosphate-5-phosphate (PRcP) to PRP (8). These reactions are summarized in Scheme 1.

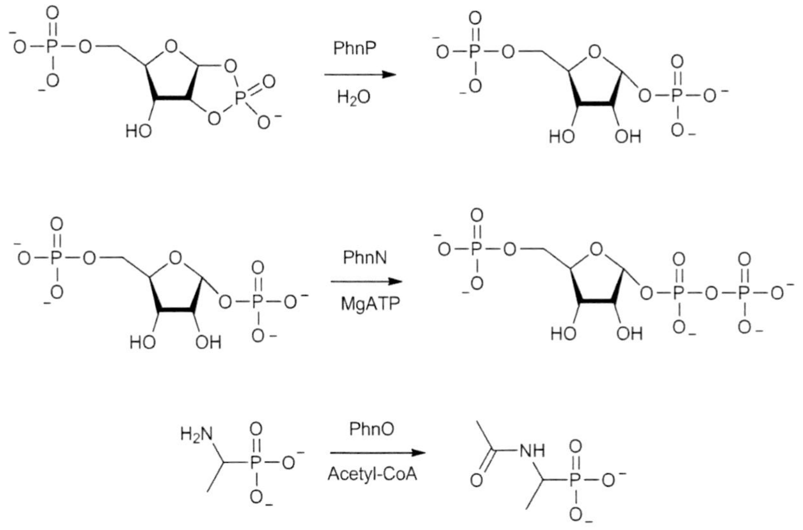

Scheme 1. Reactions catalyzed by PhnN, PhnO, and PhnP.

In *E. coli* methyl phosphonate is converted to phosphate and methane enzymatically as shown in Scheme 2 via the C-P lyase pathway. This transformation is the equivalent to the addition of water during the cleavage of the carbon-phosphorus bond, but an enzymatic mechanism for the conversion of inactivated phosphonates to phosphate is unknown.

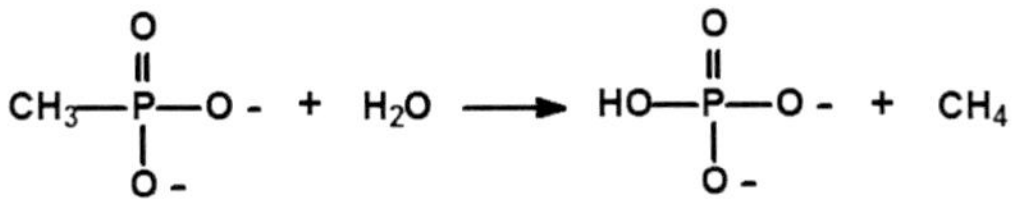

Scheme 2. Conversion of methyl phosphonate to phosphate and methane.

At the start of our investigation there were no functional annotations for any of the seven proteins considered essential for C-P bond cleavage despite three decades of research on this pathway. The primary difficulty in elucidating the reactions catalysed by these enzymes has been the extremely poor solubility of the proteins encoded by the essential gene when removed from their cellular environment. The protein encoded by *phnH* was the only protein to have been reported as soluble. This protein has been crystallized and the three-dimensional structure (PDB id: 3fsu) has been solved (9). Despite computational and ligand binding studies, no clues were available that would have provided hints of the catalytic role of PhnH in the C-P lyase pathway. PhnM was predicted to be a hydrophobic protein and this has led to the speculation that the *phnGHIJKLM* gene products form a membrane-associated enzyme complex that functions collectively to catalyse C-P bond cleavage (1, 2). Zechel, Hove-Jensen and co-workers recently purified the gene products from the expression of the genes *phnGHIJK* as a soluble multi-protein complex (2, 10). Walsh, Floss and co-workers have shown that C-P bond cleavage occurs with racemization of stereochemistry at the carbon centre. This result suggests that the C-P bond cleavage reaction proceeds via a radical-based mechanism (11). Frost and co-workers isolated α-D-ribose-1-ethylphosphonate from the extracellular medium after providing radio-labelled ethylphosphonate to *E. coli* as the sole phosphorus source (12, 13). This transformation is illustrated in Scheme 3.

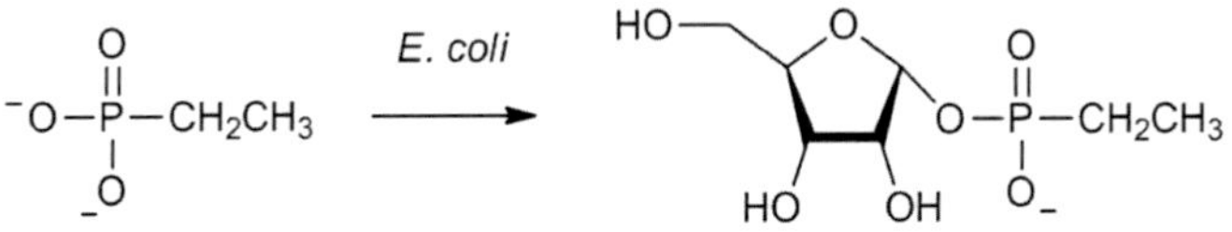

Scheme 3. Formation of α-D-ribose-1-ethylphosphonate in *E. coli* from ethylphosphonate.

Based on the functional annotations to the auxiliary proteins (PhnNOP), and the previous literature available on the C-P lyase complex in *E. coli*, there were three conclusions that could be drawn at the start of this investigation: (a) there is a ribose donor/intermediate involved in the pathway for the metabolism of phosphonates (b) the chemistry on this pathway is initiated at the C-1 position of the ribose moiety and (c) the C-P bond breaks via a radical mechanism. Sequence analysis of PhnI indicated a weak similarity to purine and pyrimidine nucleosidases at very relaxed expectation values when BLASTED using the NCBI protein database. PhnJ has four conserved cysteine residues at the C-terminal end of the protein with a highly conserved spacing between the four cysteine residues. The conserved cluster of cysteine residues is suggestive of a metal binding site and a high probability of binding an iron-sulphur cluster (14). PhnM is a member of the amidohydrolase superfamily (AHS) and is found within COG3454 of the Cluster of Orthologous Groups (COG) of proteins. Enzymes within this superfamily of proteins catalyse hydrolysis reactions at carbon and phosphorus centres using a mono- or binuclear metal centre (15). PhnK and PhnL were predicted to be ATP-binding proteins of unknown function and perhaps function as permeases (16). Therefore, the preliminary bioinformatic and sequence analysis yielded hints for three probable reactions for the metabolism of organophosphonates: (a) a nucleosidase-type reaction catalysed by PhnI – presumably involving the synthesis of a phosphonate-ribose adduct; (b) a hydrolytic reaction catalysed by PhnM; and (c) a radical-based reaction for the cleavage of the C-P bond catalysed by PhnJ that would require an iron-sulphur cluster.

PROTEIN ISOLATION

The genes for the seven essential proteins encoded by the *phnGHIJKLM* gene cluster were cloned, expressed in *E. coli* and the proteins were purified to homogeneity (17). The genes for PhnG, PhnK, PhnL and PhnM were amplified and individually ligated into a pET30a(+) vector. Of these proteins only PhnG and PhnM were soluble after cell lysis. PhnG was purified using a standard polyhistine-tag purification strategy and was stable in 150 mM NaCl at pH 8.5. PhnM was cloned without a polyhistidine-tag and was largely insoluble at pH values below 8.5. This protein was isolated using gel filtration and anion exchange chromatography at pH 9.0. Since PhnM is a member of the AHS, the growth cultures and purification buffers were supplemented with 1 mM $ZnCl_2$. The identities of PhnM and PhnG were validated by N-terminal amino acid analysis of the purified proteins. PhnK and PhnL were insoluble when expressed from a standard pET30a(+) vector. These genes were subsequently re-cloned with N-terminal glutathione S-transferase (GST) solubility tags using a pET42a(+) vector. These clones resulted in the expression of GST fused to the N-terminus of the target protein with a linker containing a Factor-Xa cleavage site. The PhnK and PhnL GST-fusion proteins were soluble and purified using a GST-affinity column (GE Healthcare). These two proteins required 1 mM dithiothreitol (DTT), 500 mM NaCl and 10% glycerol, pH 8.8, for stability (17).

A pET28b(+) vector was used for the initial cloning of the genes for PhnH, PhnI and PhnJ. Of these proteins, only PhnH was soluble after cell lysis. This protein was purified as described previously (9). Since PhnI and PhnJ were insoluble, they were re-cloned into a pET42a(+) vector with a GST-tag using the same strategy as that for PhnK and PhnL. The purification of PhnI required 1 mM EDTA, 1 mM DTT, 500 mM NaCl and 10% glycerol at pH 8.5. PhnJ required 1 mM EDTA, 500 mM NaCl and 10% glycerol at pH 8.8. Of the seven proteins critical for cleavage of phosphorus-carbon bonds, PhnG, PhnH and PhnM were purified without any solubility tags. PhnI, PhnJ, PhnK and PhnL were purified as N-terminal GST-fusion proteins. The GST-fusion proteins precipitated within ~15 minutes after removal of the N-terminal GST-tag by Factor-Xa. All protein purifications contained 100 mM HEPES buffer at the specified pH (17).

CATALYTIC FUNCTION OF PHNI

Prior investigations have hinted that a ribose intermediate is involved in the metabolism of phosphonate substrates prior to the conversion to phosphate (8, 12 – 13). Since PhnI showed a distant relationship to enzymes that are functionally validated as nucleosidases, we incubated this protein (10 µM) with a small and focused library of probable ribose donors including nucleosides, nucleotides and other ribose bearing compounds such as NAD^+, NADP and ADP-ribose at concentrations of ~2 mM for up to 6 hours in the presence of 3 mM $MgCl_2$ at pH 8.5. The liberation of the free base was determined spectrophotometrically at 240 – 350 nm in the presence of coupling enzymes (1 µM) that are capable of deaminating adenine (18), guanine (19) or cytosine (20). Xanthine oxidase (50 Units) was added to all inosine containing compounds (21). The only active substrates found in these preliminary screening assays were the purine nucleoside di- and tri- phosphates i.e. ATP, ADP, GTP, GDP, ITP and IDP. The best substrates for PhnI were GTP and ATP with kinetic constants for the nucleosidase activity of ~10^4 $M^{-1}s^{-1}$ for k_{cat}/K_m. The products of this reaction were D-ribose-5-triphosphate (RTP) and the free base (adenine or guanine when ATP or GTP were utilized, respectively). The reaction catalysed by PhnI with ATP is presented in Scheme 4. The RTP could be purified using a ResourceQ anion exchange column (GE Healthcare). The structure of RTP was initially identified by ^{31}P-NMR spectroscopy and further confirmed by one- and two-dimensional NMR spectroscopy.

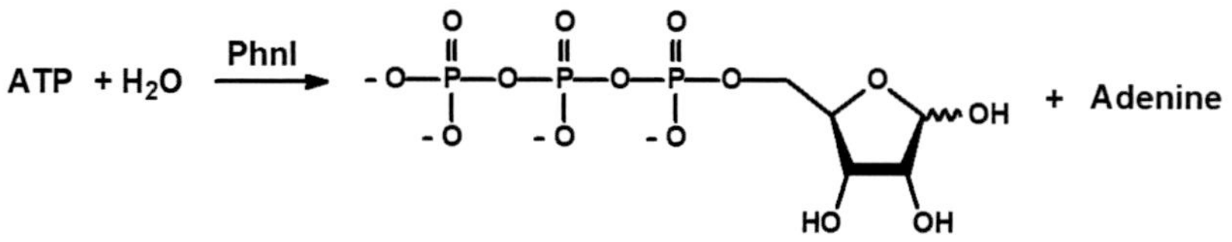

Scheme 4. Reaction catalyzed by PhnI in the presence of MgATP.

The formation of D-ribose-5-triphosphate from either ATP or GTP did not appear to be a productive route for the C-P lyase pathway since the phosphonate substrate was not connected to the ribose moiety. We therefore incubated a small library of phosphate and phosphonate derivatives (5 mM) with MgATP in the presence of PhnI to determine if any of these compounds (rather than water) could displace adenine. Surprisingly, phosphate, pyrophosphate, methylphosphate and sulfate inhibited the displacement of adenine from ATP but no inhibition was observed with methylphosphonate, ethylphosphonate, phosphonoacetic acid or 2-aminoethylphosphonate. The apparent inhibition constant for phosphate at an ATP concentration of 0.1 mM was ~0.30 mM. RTP and adenine were the only products formed in the presence of either phosphate or methylphosphonate. There were no further changes in the reaction products whether we used the GST-tagged PhnI in the presence or absence of Factor Xa (22).

Since it was previously postulated that the C-P lyase reaction involved a multi-protein complex, all of the recombinantly purified proteins were added to the reaction mixture (10). No new products were detected other than RTP and adenine. Since PhnI, PhnJ, PhnK and PhnL were all GST-fusion proteins, Factor-Xa (50 Units) was added to the reaction mixture for *in situ* cleavage of the GST-tags. With the addition of Factor-Xa, a new compound was observed by ^{31}P NMR spectroscopy at 17.8 ppm. Anion exchange chromatography was used to purify the newly formed compound. One- and two-dimensional NMR spectra were consistent with the formation of α-D-ribose-1-methylphosphonate-5-triphosphate (RPnTP) as shown in Scheme 5. A small amount of the β-anomer was also detected (approximately 5 – 10 % of the α-anomer). In this transformation, all of the ATP was consumed but the overall yield of the phosphonylated product was approximately 30 – 35 %; the rest of the ATP was converted to RTP.

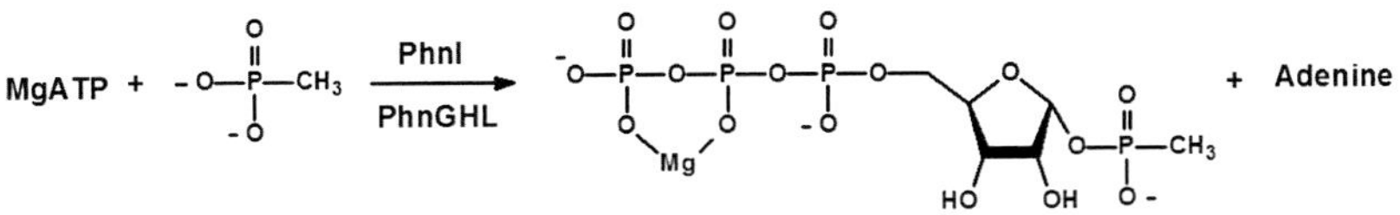

Scheme 5. Formation of RPnTP by PhnI in the presence of PhnGHL after cleavage of the GST-tags.

To determine whether all of the proteins were required for this transformation, each protein was removed individually from the reaction mixture. The minimal set of proteins required for the nucleophilic attack of methylphosphonate on the anomeric carbon of MgATP to form adenine and form RPnTP were PhnI, PhnG, PhnH and PhnL. The kinetic constants for the reaction of methylphosphonate with ATP and GTP in presence of PhnI, PhnG, PhnH and PhnL were $\sim 10^5 \, \text{M}^{-1} \text{s}^{-1}$ for k_{cat}/K_m. No reaction was observed with the fusion proteins and the GST-tags must be removed in situ for this activity. The partitioning of ATP to RTP and RPnTP likely results from partially cleaved GST-tagged protein and the precipitation of PhnI, PhnK, and PhnL after cleavage of these tags. The stoichiometry of the four proteins required for the formation of this complex and the individual functions of PhnG, PhnH and PhnL are currently uncertain.

CATALYTIC FUNCTION OF PHNM

PhnM is a member of the amidohydrolase superfamily (15). Enzymes within this super-family use a mono- or binuclear metal centre to catalyse hydrolytic reactions at carbon and phosphorus centres. PhnM is grouped within COG3454. This COG consists of approximately 200 proteins of similar sequence from different bacteria. All proteins from the amidohydrolase superfamily have a $(\beta/\alpha)_8$-barrel structural fold. Sequence alignment analysis indicated that all of the enzymes in COG3454 have a highly conserved HxD motif at the end of β-strand 1, a histidine at the end of β-strand 5, a glutamate at the end of β-strand 6 and an aspartate at the end of β-strand 8. These protein sequences lack a clearly defined ligand from β-strand 4 that could serve to bridge two divalent cations and thus PhnM was predicted to bind a single divalent metal cation in the active site at the α-metal binding site. PhnM was previously predicted to be a hydrophobic membrane bound protein (1, 2). Variation of the solution pH demonstrated that PhnM was soluble at pH values greater than 9.0. At this pH PhnM was soluble and could be purified without a solubility or affinity tag. Inductively coupled plasma mass spectrometry (ICP-MS) demonstrated that the purified PhnM contained ~ 1.2 equivalents of Zn^{2+} per monomer when Zn^{2+} was supplemented in the growth media during protein expression.

Previous studies have suggested that α-D-ribose-1-phosphonate-5-phosphate (PRPn) is a potential intermediate in the conversion of alkylphosphonates to phosphate by *E. coli* (8, 12). Hence it was rational to assume that one of the proteins in the *phn* operon would catalyse the hydrolysis of the β- and γ-phosphoryl groups from RPnTP that was isolated by the action of PhnI, PhnG, PhnH and PhnL. PhnM was the prime candidate to perform this hydrolytic reaction. In addition to a set of highly conserved residues believed to coordinate the active site divalent cation, PhnM also possess five highly conserved arginine residues on the various loops that connect the eight β-strands and α-helices. These loops are essential in determining substrate specificity. The positively charged arginine residues may participate in the coordination of the triphosphate moiety of the RPnTP substrate.

When PhnM was incubated with purified RPnTP in the presence of 1 mM $ZnCl_2$ and 5 mM $MgCl_2$, ^{31}P-NMR identified 2 new reaction products, consistent with α-D-ribose-1-methyl-phosphonate-5-phosphate (PRPn) and pyrophosphate (17). This observation was consistent with the hydrolytic cleavage of α- and β- phosphoryl groups of the triphosphate moiety of RPnTP. This reaction is shown in Scheme 6.

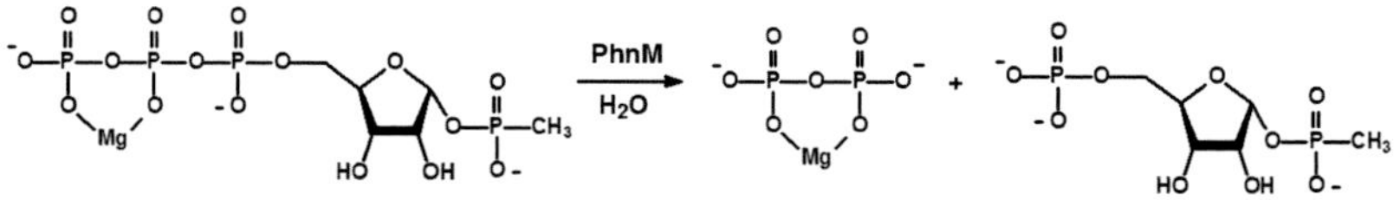

Scheme 6. Reaction catalyzed by PhnM.

To determine the exact position of the hydrolytic attack, the reaction was performed in a 1:1 mixture of H_2O and H_2O^{18}. The O^{18} label was found exclusively in the PRPn product. This result was indicative of an attack of the activated hydroxide on the α-phosphorus of the triphosphate moiety of the substrate. All of the kinetic measurements were performed on this reaction using the Pi Colorlock (Gold) phosphate detection kit from Innova Biosciences according to the manufacturer's instruction at pH 8.5. The pyrophosphate produced in this reaction was converted to inorganic phosphate by the addition of inorganic pyrophosphatase from Baker's yeast (Sigma). The kinetic constants obtained for the hydrolysis of RPnTP by PhnM were $k_{cat} = 6.4 \, s^{-1}$, $K_m = 56 \, \mu M$ and $k_{cat}/K_m = 1.1 \times 10^5 \, M^{-1}s^{-1}$. Zn^{2+} and Mg^{2+} supplementation were essential for optimal activity of PhnM.

To determine the substrate profile for PhnM, D-ribose-5-triphosphate (RTP) and D-ribose-5-diphosphate (RDP) were tried as substrates. Both RTP and RDP were synthesized by the action of PhnI on ATP and ADP, respectively, and purified using anion exchange chromatography (17). Both RTP and RDP were substrates for PhnM. In both cases D-ribose-5-phosphate was one product. The other product was pyrophosphate for RTP and inorganic phosphate for RDP. The kinetic constants for the hydrolysis of RTP by PhnM were $k_{cat} = 6.0 \, s^{-1}$, $K_m = 98 \, \mu M$ and $k_{cat}/K_m = 6.1 \times 10^4 \, M^{-1}s^{-1}$. For RDP the kinetic constants were $k_{cat} = 0.08 \, s^{-1}$, $K_m = 200 \, \mu M$ and $k_{cat}/K_m = 3.8 \times 10^2 M^{-1}s^{-1}$ (17). Thus, the substrate profiles showed that the RPnTP was the best substrate and that PhnM hydrolyses the triphosphate moiety to form the PRPn and pyrophosphate. Enzyme activity with PRPn is 300-fold more active than with RDP. The kinetic constants did not change for the hydrolysis of RPnTP when PhnG, PhnH, PhnI, PhnK, PhnL and Factor-Xa were added to the assay mixture. Therefore there is no evidence that PhnM forms a complex with these other proteins.

CATALYTIC FUNCTION OF PHNJ

The deletion of *phnJ* from *E. coli* led to the identification of α-D-ribose-1-methylphosphonate (RPn) in the growth medium and this has led to the prediction that α-D-ribose-1-methylphosphonate-5-phosphate (PRPn) is the ultimate substrate for the actual C-P lyase reaction (8). This expectation was consistent with our experimental results since this compound was synthesized from ATP and methylphosphonate by the combined actions of PhnI, PhnG, PhnH, PhnL and PhnM. Of the remaining two enzymes in the *phn* operon, the most likely protein to catalyse the actual carbon-phosphorus bond cleavage was PhnJ. PhnJ has four conserved cysteine residues with a spacing of $Cx_2Cx_{21}Cx_5C$ near the C-terminal end of the protein. These four conserved cysteine residues could form a metal-thiolate cluster or more likely, an iron-sulphur cluster. The cleavage of PRPn to PRcP (or PRP) was assumed to require a radical mechanism and such reactions could be catalysed by radical SAM enzymes.

The radical SAM superfamily was first identified in 2001 by Sofia and co-workers, based on the structural and mechanistic similarities of lysine aminomutase (LAM), biotin synthase (BioB), lipoic acid synthase (LipA) and pyruvate formate lyase activating enzyme (PflA). Each of these enzymes uses a structurally conserved $(\beta/\alpha)_8$-TIM barrel motif to produce a 5'-deoxyadenosyl radical from S-adenosyl-L-methionine (SAM) using an unusual $[Fe_4S_4]$-cluster which is redox active in the reduced $[Fe_4S_4]^{1+}$ and the oxidized $[Fe_4S_4]^{2+}$ oxidation states (23). A recent comprehensive review by Frey *et al.* (24), has suggested that more than 2800 proteins belong to this superfamily. A signature feature for the members of this superfamily is a Cx_3Cx_2C motif near the N-terminal end of the protein. These three cysteine residues nucleate the $[Fe_4S_4]$-cluster. This cluster binds SAM to the unique Fe that is not ligated to any three cysteine residues. The 5'-deoxyadenosyl radical generated from the reductive cleavage of SAM, abstracts a hydrogen atom from the substrate or the protein (to generate a protein radical) initiating the reaction. With the successful reconstitutions of ThiC (25) (Cx_2Cx_4C), HmdA (26) (Cx_5Cx_2C), and Dph2 (27), a more diverse combination of conserved cysteine residues can function for the assembly of the $[Fe_4S_4]$-cluster and perform similar radical reactions. Another comprehensive review by Booker *et al.* (28), lists more examples of members of the radical SAM superfamily lacking the canonical Cx_3Cx_2C motif as well as enzymes possessing multiple iron-sulphur clusters. Elp3, a component of the factor required for transcriptional elongation ($Cx_4Cx_9Cx_2C$) is one such example. Another example is PqqE, which is one of the six proteins required for the biosynthesis of pyrroloquinoline quinone. This protein contains a non-canonical $Cx_2Cx_{27}C$ motif binding a $[Fe_4S_4]$-cluster capable of producing a 5'-deoxyadenosyl radical upon reduction along with another cluster of the canonical Cx_3Cx_2C motif.

Literature precedents have hinted at the formation of some sort of metal-thiolate cluster on PhnJ (14), based on the highly invariant $Cx_2Cx_{21}Cx_5C$ motif at the C-terminal of the protein. PhnJ purified aerobically with a GST-tag, had a black colour. This was presumably due the non-specific binding of iron sulphide from the degradation of an iron-sulphur cluster upon exposure to air. Metal analysis using ICP-MS showed the presence of ~2.2 equivalents of Fe per monomer of aerobically purified PhnJ. Thus PhnJ appeared to be capable of binding iron and sulphide. The metal ions bound to the aerobically purified PhnJ were removed with the addition of EDTA to the purification buffers. The apo-PhnJ was made anaerobic by passing argon through the protein solution and then an excess of sodium dithionite was added (5 mM). PhnJ was incubated in the glove box for four hours. Through-out the reconstitution experiments the oxygen concentration was maintained below 4 ppm. Reconstitution of the $[Fe_4S_4]$-cluster was initiated by slow dropwise addition of a 50-fold enzyme excess of $FeSO_4$. After equilibrating anaerobically for three hours, a 50-fold excess of Na_2S was added and the mixture incubated for an additional three hours. The excess iron and sulphide that formed insoluble iron sulphide (black solid) was removed by centrifuga-tion, followed by ultrafiltration through a 10 kDa membrane. The protein with the recon-stituted iron-sulphur cluster had a reddish-brown colour. The UV-visible absorption spec-trum of PhnJ, reconstituted with iron and sulphide, had a broad absorption band centred at 403 nm that is indicative of a $[Fe_4S_4]$-cluster (23, 24, 28). The absorption band disappears upon addition of 1 mM dithionite, suggesting the reduction of the $[Fe_4S_4]^{2+}$ cluster to the $[Fe_4S_4]^{1+}$ species, which is the active form for most radical SAM enzymes (17). The discoloration is also indicative of the redox activity of the proteins within the radical SAM superfamily (24, 28). Quantitation of the UV-visible spectrum of PhnJ suggested that the chemical reconstitution yield was approximately 40%.

Even though there were prior suggestions that PRPn was the ultimate substrate for PhnJ, the putative C-P bond cleaving enzyme from the gene knockout studies carried out by Zechel, Hove-Jensen and co-workers (8, 10), there were still two compounds that could have been potential substrates for cleavage by PhnJ of the C-P bond: PRPn and RPnTP. Both of these compounds were incubated with 125 µM PhnJ that was chemically reconstituted with a $[Fe_4S_4]$-cluster, 2 mM SAM and 1mM dithionite, under anaerobic conditions at pH 6.8. The reactions were analysed by ^{31}P NMR spectroscopy but there was no change in the NMR spectrum for either substrate. Since PhnJ was expressed and purified as an N-terminal GST-fusion protein, the reaction was supplemented with Factor-Xa to cleave the GST-tag. No change in the ^{31}P NMR spectrum was observed with RPnTP, but a new resonance appeared at 16.2 ppm using PRPn as a substrate (17). The increase in the resonance at 16.2 ppm correlated with a decrease in the phosphonate resonance of PRPn at 16.6 ppm. The new resonance splits into a triplet in the ^1H-coupled ^{31}P spectrum (17) demonstrating that the product is no longer a methyl phosphonate. The new resonance was consistent with a cyclic phosphate and the proton coupling constant of 21 Hz indicates that the phosphate moiety of the product was esterified to the hydroxyls attached to C 1 and C 2 of the ribose. This result was consistent with the findings of Zechel, Hove-Jensen and co-workers for the substrate of PhnP (8, 17). Thus, PhnJ requires a reduced $[Fe_4S_4]$-cluster and SAM to catalyse the formation of α-D-ribose-1,2-cyclic phosphate-5-phosphate from PRPn. The overall reaction is illustrated in Scheme 7.

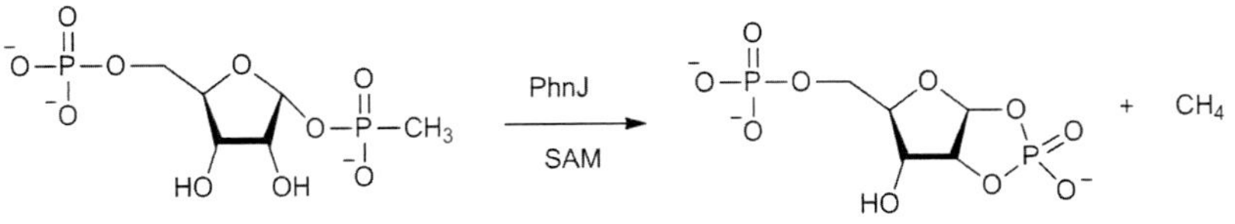

Scheme 7. Reaction catalyzed by PhnJ in the presence of SAM.

Gas chromatography (GC) and GC-mass spectrometry (MS) were used to confirm the formation of methane. All GC and GC-MS samples were prepared anaerobically. The reconstituted PhnJ was incubated with 2 mM SAM, 1 mM dithionite, 5 mM PRPn, and Factor-Xa in a sealed tube for 5 hours. 50 μL of the headspace (500 μL) was removed using a gas-tight Hamilton syringe and injected into the GC instrument (Hewlett Packard HP 6890 GC system with FID detector and manual injection) equipped with a 30 m X 0.32 mm I.D. SGE Solgel 1 column used in splitless injection mode, at 20 psig He carrier pressure, with a temperature program of 50 °C/1 min, to 100 °C at 10 °C/min. The same reaction was conducted for the GC-MS experiment using a Thermo Electron Corporation DSQ quadrupole GC-MS instrument with Finnigan Trace GC Ultra chromatograph at 70 E.V. EI ionization. The temperature program and injection mode were the same. The column was a 30 m x 0.25 mm I.D. SGE BP1 column operated at 1 mL/min constant flow of He.100 μL of the headspace was injected into the GC-MS instrument. Gas chromatographic analysis of the headspace above the liquid showed a single peak that co-eluted with a methane standard. The formation of methane was confirmed by coupling the output of the GC to a mass spectrometer and detection of a mass of 16. Thus, the two products formed from the action of PhnJ on PRPn are methane and PRcP. To determine whether there were any byproducts of this reaction, an analysis was conducted to check for the formation of methanol and formaldehyde. Alcohol dehydrogenase from yeast was used to test the formation of methanol using NAD^+. The reaction showed no formation of methanol. For the detection of formaldehyde, purpald was used. A standard curve was made and the reaction mixture showed no formation of formaldehyde. The addition of SAM, PhnJ, a reductant (dithionite) and Factor-Xa were required for the formation of PRcP and methane from the $[Fe_4S_4]^{1+}$ reconstituted PhnJ under strictly anaerobic conditions.

To determine the fate of SAM during the reaction, HPLC and amino acid analysis were employed. Amino acid analysis of the reaction products from SAM in the reaction catalysed by PhnJ was performed by the Protein Chemistry Laboratory, Texas A&M University. The derivatized amino acids were separated by reverse phase HPLC and detected by UV absorbance with a diode array detector or by fluorescence using an in-line fluorescence detector. For the reactions involving PhnJ, the concentrations of the components were as follows:

PhnJ = 70 μM, SAM = 2.0 mM, dithionite = 1.0 mM, PRPn = 1.0 mM in 150 mM HEPES, containing 250 mM NaCl, 10% v/v glycerol and 1X Factor Xa reaction buffer at pH 8.5 in a volume of 1.0 mL. The reaction was initiated by adding 50 units of Factor Xa. As a control, PhnJ was eliminated from the reaction mixture. All of the reactions were kept anaerobic for three hours after which the precipitated enzyme was removed by centrifugation. The reaction was filtered through a 3 kDa ultra-filtration membrane and the flow-through was collected. The control reaction showed less than 0.01 μM L-methionine. In the reaction mixture with PhnJ, 150 μM of L-methionine was detected. The HPLC analysis for determining the reaction products of SAM with PhnJ was performed with an AKTA Purifier FLPC/HPLC system with a C18 column (Cosmosil 5C18-AR-II 4.6 × 150 mm Nacalai, USA). The C18-column was pre-calibrated with SAM, 5'-deoxyadenosine (5DA) and 5'-deoxy-5'-methylthioadenosine (MTA) for the elution profiles of the standards. For the reactions involving PhnJ, the concentrations of the components were as follows: PhnJ = 60 μM, SAM = 0.5 mM, dithionite = 1 mM, α-D-ribose-1-methylphosphonate-5-phosphate = 2 mM in 150 mM HEPES, containing 250 mM NaCl, 10% v/v glycerol and Factor-Xa reaction buffer at pH 8.5 in a volume of 1.0 mL. The reaction was initiated by adding 50 units of Factor-Xa to the reaction. As a control, dithionite was eliminated in one reaction and PhnJ was eliminated in another. All of the reactions were kept anaerobic for 3 hours after which the precipitated enzyme was removed by centrifugation. The reaction was filtered through a 3 kDa ultra-filtration membrane and the flow-through was collected. 250 μL of this material was loaded onto a C18-column and the HPLC traces were collected. HPLC analysis showed the formation of 5'-deoxyadenosine and amino acid analysis confirmed the formation of methionine. These products were only formed when PhnJ (with a reconstituted iron-sulphur centre), Factor-Xa, SAM, dithionite and PRPn were added to the reaction mixture under strictly anaerobic conditions. The omission any one of the above components resulted in no formation of 5'-deoxyadenosine. The concentration of PRcP, 5'-deoxyadenosine, methionine, and methane formed in the presence of PhnJ showed that 1 – 10 turnovers of product were formed per PhnJ. The small number of turnovers per enzyme may reflect the poor solubility of PhnJ after proteolytic cleavage of the N-terminal GST-fusion tag.

RECONSTITUTION OF THE C-P LYASE PATHWAY

The mechanism for the conversion of alkyl phosphonates to phosphate in *E. coli* was elucidated. PhnI (a novel nucleosidase), in the presence of PhnG, PhnH, and PhnL, catalyses the formation of RPnTP from MgATP and methylphosphonate. PhnM (an amidohydrolase) catalyses the hydrolysis of RPnTP to form pyrophosphate and PRPn. PhnJ can be chemically reconstituted with a $[Fe_4S_4]$-cluster with ferrous sulfate, sodium sulphide and sodium dithionite under strictly anaerobic conditions. The reconstituted PhnJ catalyses the homolytic cleavage of the C-P bond of PRPn to form PRcP and methane in the presence of SAM. These reactions are summarized in Scheme 8.

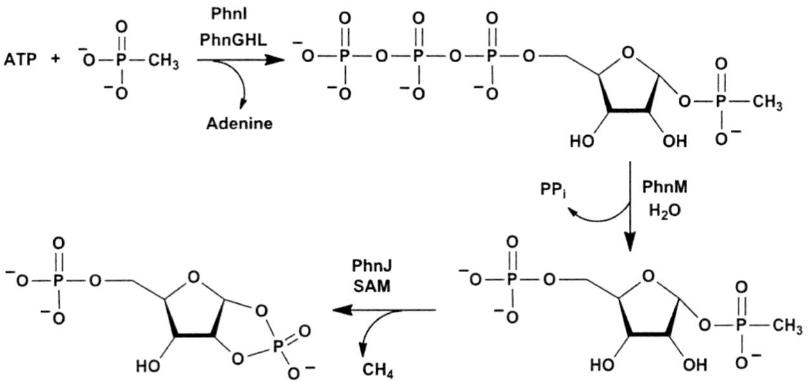

Scheme 8. Reaction pathway for the conversion of methyl phosphonate to methane and PRcP.

PhnJ is a putative radical SAM enzyme and the working model for the cleavage of the C-P bond of the phosphonate moiety is presented in Scheme 9 (17). In this mechanism the reduced iron-sulphur cluster transfers an electron to SAM and forms an oxidized iron-sulphur cluster, methionine, and 5'-deoxyadenosyl radical. The 5'-deoxyadenosyl radical abstracts a hydrogen atom from the protein generating a protein based radical and to generate 5'-deoxyadenosine. This protein based radical attacks the phosphorus of the phosphonate forming a protein-substrate adduct. This results in the homolytic cleavage of the C-P bond, coupled with formation of a transient methyl radical. Methane is formed presumably from the abstraction of a hydrogen atom from either the protein residue to regenerate the protein radical or from 5'-deoxyadenosine to regenerate the 5'-deoxyadenosyl radical. Attack of the C 2 hydroxyl generates the cyclic diester phosphate product PRcP. The formation of PRcP by PhnJ is consistent with the known function of PhnP that catalyses the hydrolysis of PRcP to PRP (8).

$$[Fe_4S_4]^{1+} + SAM \rightleftharpoons [Fe_4S_4]^{2+} + L\text{-}Met + Ado\text{-}\dot{C}H_2$$

$$Ado\text{-}\dot{C}H_2 + PhnJ\text{-}XH \rightleftharpoons Ado\text{-}CH_3 + PhnJ\text{-}X\cdot$$

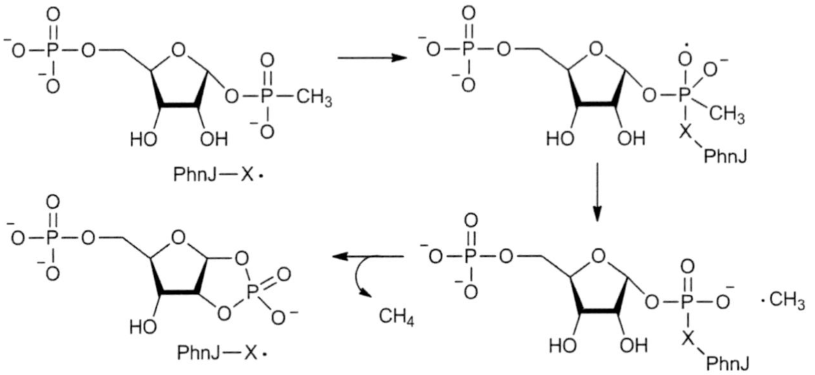

Scheme 9. Working model for the reaction mechanism catalyzed by PhnJ.

REFERENCES

[1]   Metcalf, W.W. and Wanner, B.L. (1993) Evidence for a fourteen-gene, *phnC* to *phnP* locus for phosphonate metabolism in *Escherichia coli*. *Gene* **129**:27 – 32. doi: 10.1016/0378-1119(93)90692-V.

[2]   Metcalf, W.W. and Wanner, B.L. (1993) Mutational analysis of an *Escherichia coli* fourteen-gene operon for phosphonate degradation, using Tnpho*A*' elements. *J. Bacteriol.* **175**:3430 – 3442.

[3]   Ternan, N.G., McGrath, J.W., McMullan, G. and Quinn, J.P. (1998) Organophosphates: occurrence, synthesis and biodegradation by microorganisms. *World J. Microbiol. Biotechnol.* **14**:635 – 647. doi: 10.1023/A:1008848401799.

[4]   Kononova, S.V., and Nesmeyanova, M.A. (2002) Phosphonates and their degradation by microorganisms. *Biochemistry (Moscow)* **67**:184 – 195. doi: 10.1023/A:1014409929875.

[5]   White, A.K. and Metcalf, W. (2007) Microbial metabolism of reduced phosphorus compounds. *Annu. Rev. Microbiol.* **61**:379 – 400. doi: 10.1146/annurev.micro.61.080706.093357.

[6]   Hove-Jensen, B., Rosenkrantz, T.J., Haldimann, A., and Wanner, B.L. (2003) *Escherichia coli phnN*, encoding ribose 1,5-bisphosphokinase activity (phosphoribosyl diphosphate forming): dual role in phosphonate degradation and NAD biosynthesis pathways. *J. Bacteriol.* **185**:2793 – 2801.
doi: 10.1128/JB.185.9.2793-2801.2003.

[7]   Errey, J.C. and Blanchard, J.S. (2006) Functional annotation and kinetic characterization of PhnO from *Salmonella enterica. Biochemistry* **45**:3033 – 3039.
doi: 10.1021/bi052297p.

[8]   Hove-Jensen, B., McSorley, F.R. and Zechel, D.L. (2011) Physiological role of *phnP*-specified phosphoribosyl cyclic phosphodiesterase in catabolism of organophosphoric acids by the carbon-phosphorus lyase pathway. *J. Am. Chem. Soc.* **133**:3617 – 3624.
doi: 10.1021/ja1102713.

[9]   Adams, M.A., Luo, Y., Hove-Jensen, B., He, S.M., van Staalduinen, L.M., Zechel, D.L., and Jia, Z. (2008) Crystal structure of PhnH: an essential component of the carbon-phosphorus lyase in *Escherichia coli. J. Bacteriol.* **190**:1072 – 1083.
doi: 10.1128/JB.01274-07.

[10]   Jochimsen, B., Lolle, S., McSorley, F.R., Nabi, M., Stougaard, J., Zechel, D.L., and Hove-Jensen, B. (2011) Five phosphonate operon gene products as components of a multi-subunit complex of the carbon-phosphorus lyase pathway. *Proc. Nat. Acad. Sci. U.S.A.* **108**:11393 – 11398.
doi: 10.1073/pnas.1104922108.

[11]   Ahn, Y., Ye, Q., Cho, H., Walsh, C.T. and Floss, H.G. (1992) Stereochemistry of carbon-phosphorus cleavage in ethylphosphonate catalysed by C-P lyase from *Escherichia coli. J. Am. Chem. Soc.* **114**:7953 – 7954.
doi: 10.1021/ja00046a075.

[12]   Avila, L.Z., Draths, K.M., and Frost, J.W. (1991) Metabolites associated with organophosphonate C-P bond cleavage: chemical synthesis and microbial degradation of [^{32}P]-ethylphosphonic acid. *Bioorg. & Med. Chem. Lett.* **1**:51 – 54.
doi: 10.1016/S0960-894X(01)81089-1.

[13]   Frost, J.W., Loo, S., Cordeiro, M.L., and Li, D. (1987) Radical-based dephosphorylation and organophosphonate biodegradation. *J. Am. Chem. Soc.* **109**:2166 – 2171.
doi: 10.1021/ja00241a039.

[14]   Parker, G.F., Higgins, T.P., Hawkes, T., and Robson, R.L. (1999) *Rhizobium (Sinorhizobium) meliloti phn* genes: characterization and identification of their protein products. *J. Bacteriol.* **181**:389 – 395.

[15] Seibert, C.M., and Raushel, F.M. (2005) Structural and catalytic diversity within the amidohydrolase superfamily. *Biochemistry* **44:**6383 – 6391.
doi: 10.1021/bi047326v.

[16] Hartley, L.E., Kaakoush, N.O., Ford, J.L., Korolik, V. and Mendz, G.L. (2009) Characterization of *Campylobacter jejuni* genes potentially involved in phosphonate degradation. *Gut Pathogens* **1:**13.
doi: 10.1186/1757-4749-1-13.

[17] Kamat, S.S., Williams, H.J., and Raushel, F.M. (2011) Intermediates in the transformation of phosphonates to phosphate by bacteria. *Nature* **480**(7378):570 – 573.
doi: 10.1038/nature10622.

[18] Kamat, S.S., Bagaria, A., Kumaran, D., Holmes-Hampton, G.P., Fan, H., Sali, A., Sauder, J.M., Burley, S.K., Lindahl, P.A., Swaminathan, S. and Raushel, F.M. (2011) Catalytic mechanism and three-dimensional structure of adenine deaminase. *Biochemistry* **50:**1917 – 1927.
doi: 10.1021/bi101788n.

[19] Maynes, J.T., Yuan, R.G. and Snyder, F.F. (2000) Identification, expression and characterization of *Escherichia coli* guanine deaminase. *J. Bacteriol.* **182:**4658 – 4660.
doi: 10.1128/JB.182.16.4658-4660.2000.

[20] Hall, R.S., Fedorov, A.A., Xu, C., Fedorov, E.V., Almo, S.C., and Raushel, F.M. (2011) Three dimensional structure and catalytic mechanism of cytosine deaminase. *Biochemistry* **50:**5077 – 5085.
doi: 10.1021/bi200483k.

[21] Krenitsky, T.A., Neil, S.M., Elion, G.B. and Hitchings, G.H. (1972) A comparison of the specificities of xanthine oxidase and aldehyde oxidase. *Arch. Biochem. Biophy.* **150:**585 – 599.
doi: 10.1016/0003-9861(72)90078-1.

[22] La. Vallie, E.R., McCoy, J.M., Smith, D.B. and Riggs, P. (1994) Enzymatic and chemical cleavage of fusion proteins. *Curr. Protocols Mol. Biol.* Unit 16.4B.

[23] Sofia, H.J., Chen, G., Hetzler, B.G., Reyes-Spindola, J.F. and Miller, N. E. (2001) Radical SAM, a novel protein superfamily linking unsolved steps in familiar biosynthetic pathways with radical mechanisms: functional characterization using new analysis and information visualization methods. *Nucleic Acids Research* **29**(5):1097 – 1106.
doi: 10.1093/nar/29.5.1097.

[24]   Frey, P.A., Hegeman, A.D., and Ruzicka, F.J. (2008) The radical SAM superfamily. *Crit. Rev. Biochem. Mol. Biol.* **43**:63 – 88. doi: 10.1080/10409230701829169.

[25]   Chatterjee, A., Li, Y., Zhang, Y., Grove, T.L., Lee, M., Krebs, C., Booker, S.J., Begley, T. P. and Ealick, S. E. (2008). Reconstitution of ThiC in thiamine pyrimidine biosynthesis expands the radical SAM superfamily. *Nature Chem. Biol.* **4**:758 – 765. doi: 10.1038/nchembio.121.

[26]   McGlynn, S.E., Boyd, E.S., Shepard, E.M., Lange, R.K., Gerlach, R., Broderick, J.B. and Peters, J.W. (2010). Identification and characterization of a novel member of the radical AdoMet enzyme superfamily and implications for the biosynthesis of the Hmd hydrogenase active site cofactor. *J. Bacteriol.* **192**:595 – 598. doi: 10.1128/JB.01125-09.

[27]   Zhang, Y., Zhu, X., Torelli, A.T., Lee, M., Koralewski, R.M., Wang, E., Freed, J., Krebs, C., Ealick, S.E., and Lin, H. (2010) Diphthamide biosynthesis requires an organic radical generated by iron-sulphur enzyme. *Nature.* **465**:891 – 896. doi: 10.1038/nature09138.

[28]   Booker, S.J., and Grove, T.L. (2010) Mechanistic and functional versatility of radical SAM enzymes. *F1000 Biol. Rep.* **2**:52.

Beilstein-Institut

Experimental Standard Conditions of Enzyme Characterization
September 12th – 16th, 2011, Rüdesheim/Rhein, Germany

19

IMP Dehydrogenase:
the Dynamics of Reaction Specificity

Lizbeth Hedstrom

Departments of Biology and Chemistry, Brandeis University, 415 South St.,
Waltham MA 02454, U.S.A.

E-Mail: hedstrom@brandeis.edu

Received: 27th February 2012/Published: 15th February 2013

Abstract

Subtle changes in enzyme structure can have enormous impact on catalysis, as vividly illustrated in IMP dehydrogenase (IMPDH) and GMP reductase (GMPR). These proteins share a common structure and set of catalytic residues and bind the same ligands with similar affinities. IMPDH catalyses a hydride transfer reaction involving a nicotinamide cofactor, with formation of the covalent intermediate E-XMP*. Hydrolysis of this intermediate produces XMP, which is converted to GMP by the action of another enzyme. In the GMPR reaction, E-XMP* is formed by the deamination of GMP, and is subsequently reduced via a hydride transfer reaction with a nicotinamide cofactor. In both cases, a conformational change separates the two chemical transformations. The protein moves in the case of IMPDH, while the cofactor moves in GMPR. Thus conformational dynamics control reaction specificity in the IMPDH/GMPR family, with intriguing implications for the evolution of these enzymes.

Introduction

The remarkable versatility of the $(\beta/\alpha)_8$ barrel fold, also known as the TIM barrel, is well recognized [1 – 6]. The current SCOP and CATH databases list approximately thirty $(\beta/\alpha)_8$ barrel protein superfamilies [7, 8], which catalyse over twenty-five different reactions [9]. One $(\beta/\alpha)_8$ barrel protein superfamily contains just two proteins, inosine monophosphate dehydrogenase (IMPDH) and guanosine monophosphate reductase (GMPR), which catalyse similar reactions with opposing metabolic consequences (Figure 1; [10]). IMPDH catalyses the oxidation of IMP to XMP with concomitant reduction of NAD$^+$. This reaction is the first committed and rate-limiting step in the biosynthesis of guanine nucleotides, and therefore controls the size of the guanine nucleotide pool. IMPDH inhibitors block cell proliferation, which makes IMPDH an attractive drug target. IMPDH inhibitors are currently used as immunosuppressive [11], anticancer [12, 13] and antiviral therapy [14], and also show promise as antimicrobial agents [15]. GMPR catalyses the reduction of GMP to IMP and ammonia with the concomitant oxidation of NADPH. This reaction allows guanine nucleotides to be recycled into the adenine nucleotide pool. These two enzymes bind the same ligands and have the same catalytic residues. The same covalent intermediate E-XMP* is formed during both reactions. Recent work provides some insights into why this intermediate partitions in different directions on the two enzymes.

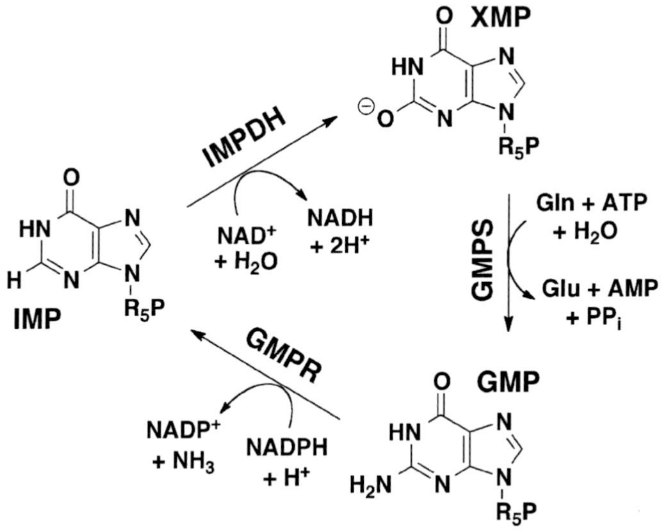

Figure 1. The interconversions of IMP and GMP.

THE IMPDH REACTION

IMPDH catalyses two distinct chemical transformations. In the first step, Cys319 attacks C2 of IMP and hydride is transferred to NAD$^+$ to produce the covalent intermediate E-XMP* (Figure 2; *Tritrichomonas foetus* IMPDH numbering will be used throughout). This step is likely to proceed via a tetrahedral intermediate, although experiments have yet to address this point. The hydrolysis of E-XMP* produces XMP; this step is rate-limiting in most IMPDHs [10, 16].

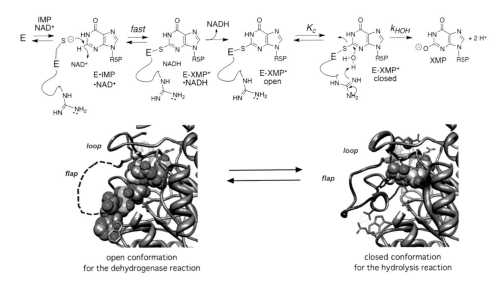

Figure 2. The mechanism of IMPDH. The model for the open conformation is the E•IMP•tiazofurin adenine dinucleotide complex (1lrt; [36]) and the model for the closed conformation is the E•MZP complex (1pvn; [17]).

The central challenge of the IMPDH reaction is "how can a single active site catalyse two different chemical transformations?" IMPDH solves this problem with a conformational change that remodels part of the active site, as revealed by the structure of the IMPDH•mizoribine monophosphate (MZP) complex [17]. Unlike other IMP analogues, the affinity of MZP negatively correlates with the activity of mutant enzymes [18], which is behaviour characteristic of transition state analogues. The structure of the E•MZP complex does indeed resemble a transition state [17]. Cys319 and a putative catalytic water molecule are positioned in a tetrahedral arrangement as expected for the hydrolysis of E-XMP*. If E-XMP* is modelled in place of MZP, the water is poised for attack at the C2 position (Figure 3). A mobile flap (residues 412–432), disordered in previous IMPDH structures, is found in a "closed conformation", and folded into the cofactor site. The catalytic water molecule forms hydrogen bonds to Thr321 from the same loop that contains the catalytic Cys319, as well as two residues of the flap, Arg418 and Tyr419 (Figure 3). Thr, Arg and Tyr residues are not usually candidates for general base catalysts, though precedence exists for each (see examples in [19]). The substitution of Thr321 decreases both the hydride transfer and hydrolysis steps by a factor of ~20 [20]. In contrast, mutations of both Arg418 and Tyr419 selectively perturb the hydrolysis step (Table 1). However, only the substitution of Arg418 decreases the rate of the hydrolysis step by the magnitude expected for loss of a general base [21]. As expected, the mutation of Arg418 also perturbs the equilibrium between the open and closed conformations of the flap, but the magnitude of this effect is not sufficient to account for the decrease in k_{cat} (Table 1). Guanidinium analogues rescue the activity of Arg419 mutants and Bronsted analysis of the rescue reaction suggests that the proton is almost completely transferred in the transition state, providing further support for the role of Arg418 as the base that activates water [22].

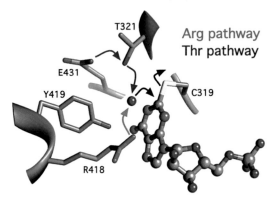

Figure 3. The activation of water in IMPDH. Model of the E-XMP*$_{closed}$ complex based on the structures 1pvn and 1jr1. The Arg pathway for water activation is depicted in teal and the Thr pathway for water activation is shown in maroon.

Table 1. Effects of mutations of the residues that interact with the catalytic water a. [21]; b. [20]

Variant	k_{cat} (s^{-1})	Hydride transfer (s^{-1})	NADH release (s^{-1})	K_c	k_{HOH} (s^{-1})
Wild-type [a]	1.9	93	8.5	140	4
Thr321Ala [b]	0.18	1.7	≥8	≥20	0.18
Arg418Ala [a]	0.004	42	11	1	0.008
Arg418Gln [b]	0.0069	≥400	≥4	10–50	0.007
Arg418Lys [b]	0.15	83	6.5	≤0.1	≥1
Tyr419Phe [a]	0.22	70	10	20	0.22

While all of the biochemical experiments are consistent with Arg418 playing the role of general base in the activation of water in the IMPDH reaction, none are definitive. Therefore, we performed combined molecular mechanics/quantum mechanics simulations to gain further insight into this transformation [23]. In the lowest energy pathway for the hydrolysis of E-XMP*, neutral Arg418 abstracts a proton from the catalytic water as it attacks the C2 position of IMP (Figure 3). Assuming that the pK$_a$ of Arg418 is ~12.5, the calculated energy barrier is in good agreement with the barrier observed experimentally. Proton transfer is rate-limiting and almost complete in the transition state, also consistent with experimental findings. More importantly, the simulations also revealed a second pathway for the activation of water that operates when Arg418 is protonated: Thr321 activates water via a proton relay with Glu431 (Figure 3C). This pathway is also consistent with experimental findings. The energy barrier is similar to that observed when Arg418 is mutated. Proton transfer is rate limiting in the simulations, and a larger solvent isotope effect is observed when Arg418 is mutated, consistent with the transfer of two protons. These simulations suggest that the Thr321 pathway activates water at low pH, while the Arg418 pathway dominates at high pH, which further predicts that the substitution of Glu341 with Gln will disable the Thr321 pathway and shift the pH rate profile to the right. This behaviour is in fact observed experimentally. Thus water appears to be activated by both the Arg418 pathway and Thr321 pathway.

Why would IMPDH possess two sets of catalytic machinery to activate water? We hypothesize that the Thr321 pathway is a remnant of evolution. Phylogenetic analysis indicates that the closest relative to IMPDH is GMPR, and that the ancestral IMPDH/GMPR contained the catalytic Cys, Thr and Glu residues [23]. IMPDH obtained the more efficient Arg418 pathway with the attendant conformational change later, at which point the Thr321 pathway became redundant. Indeed, the mutation of Glu341 to Gln has little effect on the IMPDH reaction other than shifting the pH rate profile, and many modern IMPDHs contain this substitution. Thus the simulations provided unexpected insight into the evolution of IMPDH.

THE GMPR REACTION

The GMPR reaction is effectively the reverse of the IMPDH reaction, reducing GMP to IMP with the release of ammonia. Thus GMPR also must perform two chemical transformations, a deamination reaction and a hydride transfer. Comparatively little was known about the mechanism of GMPR beyond steady-state kinetic experiments that indicated that the E•GMP•NADPH ternary complex must form before the reaction proceeds in the human enzyme [24, 25]. We and others recently confirmed that a similar pattern is observed in the reaction of *Escherichia coli* GMPR [26, 27]. Surprisingly, an isotope effect is observed on hydride transfer, indicating that hydride transfer is rate-limiting [26, 27]. These observations indicate that the cofactor is present throughout the entire reaction cycle, so the GMPR reaction cannot simply follow a reaction sequence that is the reverse of IMPDH.

If the above phylogenetic analysis is correct, then the analogous Cys186, Thr188 and Glu289 residues should be necessary for GMPR catalysis (*E. coli* GMPR numbering). Moreover, these residues should perform similar roles in the two reactions, that is, Cys186 attacks C2 to form E-XMP*, with Thr188 and Glu289 activating the leaving ammonia. The formation of E-XMP* from GMP was demonstrated by radiolabeling and mass spectroscopy, but this intermediate only forms in the presence of cofactor [26]. E-XMP* also forms from 2-Cl-IMP, but again only in the presence of cofactor. Curiously, E-XMP* forms in the presence of $NADP^+$ in addition to NADPH, which indicates that the cofactor induces a protein conformational change required for the deamination reaction. The structure of E•GMP (PDB accession number 1ble) reveals that the Cys186 and Glu289 point away from the active site and Arg286 folds across the face of the guanine ring, further protecting GMP from reaction (Figure 4; [28]). As found in the structure of the E•IMP•NADPH (PDB accession number 2c6q), when the cofactor binds, Arg286 forms part of the 2'-phosphate binding site. Movement of this segment allows Cys186, Thr188 and Glu289 to rearrange into catalytically competent alignment (Figure 4; [26]). Surprisingly, this crystal structure finds the cofactor in two conformations: (1) the 'in' conformation, where the nicotinamide is stacked against the hypoxanthine ring, optimally aligned for hydride transfer (Figure 4) and (2) the 'out' conformation, where the nicotinamide is too far from the hypoxanthine ring for hydride transfer to occur (Figure 4). An electron density, modelled as water, is observed within hydrogen bonding distance of Thr188 and the carboxamide of the cofactor, in a similar position to the catalytic water of the closed conformation of IMPDH. Therefore we hypothesize that the cofactor moves to the 'out' conformation for the deamination reaction. Further, these observations suggest that cofactor amide may be part of the machinery that activates the amine leaving group.

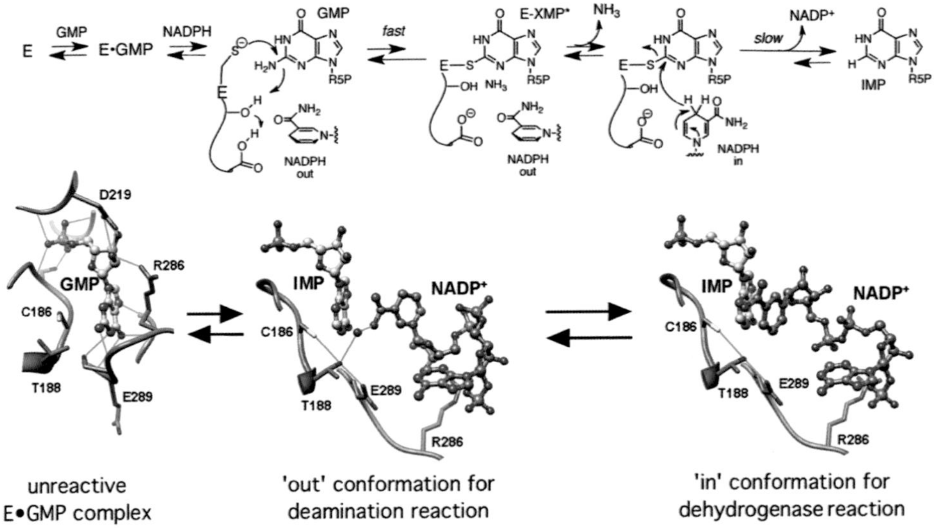

Figure 4. The mechanism of the GMR reaction. Structure of the inactive E•GMP complex is 2a7r [28]. Structure of E•IMP•NADPH in the "out" position is subunit E from 1c6q; [26]. Structure of E•IMP•NADPH in the "in" position for is subunit C from 1c6q[26].

We examined the reactions of GMPR with 2-Cl-IMP and reduced acetyl pyridine adenosine dinucleotide phosphate (APADPH) to further probe the mechanism of the deamination reaction (Table 2). The value of k_{cat} for the reaction of 2-Cl-IMP and NADPH is equivalent to that of the physiological reaction with GMP. Importantly, chloride is a good leaving group, so the 2-Cl-IMP reaction should proceed even if the residues that activate the leaving group of GMP have been removed. As expected, substitution of Cys186 with Ala completely blocks both the reactions of GMP and 2-Cl-IMP [26]. The mutation of Thr188 decreases the value of k_{cat} for the GMP reaction by a factor of 500, but reduces that for the 2-Cl-IMP reaction by only a factor of 14. This observation indicates that Thr186 is part of the machinery that activates the amine leaving group of GMP. Likewise, mutation of Glu289 also decreases the value of k_{cat} for the GMP reaction by a factor of 103, but reduces that for the 2-Cl-IMP reaction by only a factor of 20, again consistent with the involvement of this residue in leaving group activation. In APADPH, the cofactor amide is replaced with a methyl ketone. This analogue cannot catalyse the reduction of GMP ($V_{max} \leq 0.6\%$ of the reaction with NADPH), but can catalyse the reaction with 2-Cl-IMP ($V_{max} = 20\%$ of the reaction with NADPH). Thus the cofactor amide is part of the catalytic machinery activating the amine leaving group of GMP. We believe that this may be an unprecedented role for a nicotinamide cofactor. These experiments suggest that GMPR uses a complementary reaction strategy to that of IMPDH, with the cofactor moving to accommodate both chemical transformations.

Table 2. Reactions of GMPR variants. Data from [26].

Reaction	k_{cat} (s^{-1})			
	wild-type	C186A	T188A	E289Q
GMP + NADPH	0.35 ± 0.01	≤ 0.0001	(8.8 ± 0.6) x 10^{-4}	(3.8 ± 0.3) x 10^{-4}
2-Cl-IMP + NADPH	0.40 ± 0.01	≤ 0.0001	0.021 ± 0.001	0.027 ± 0.001
GMP + APADPH	< 0.0008	n.d.	n.d.	n.d.
2-Cl-IMP + APADPH	0.08 ± 0.01	n.d.	n.d.	n.d.

Can a similar cofactor conformational change occur on IMPDH? The cofactor conformational change requires a rotation about the 5' carbon of the adenosine ribose of NADPH (Figure 4). The NAD binding site of IMPDH appears to be incompatible with this motion. The cofactor occupies a different portion of the barrel domain than in GMPR in the structure of IMPDH•cofactor complexes that are currently available. This observation suggests that the adenosine binding portion of the cofactor binding site is a critical determinant of reaction specificity. Curiously, some prokaryotic IMPDHs may bind NAD$^+$ in the same manner that GMPR binds NADPH. A recent structure of *Cryptosporidium parvum* IMPDH finds an inhibitor binding in a site that resembles the adenosine subsite of GMPR [29]. The inhibitor interacts with a Tyr residue in the neighbouring subunit, much as the adenosine of NADP$^+$ does in GMPR (Figure 5). Many bacterial IMPDHs share this motif [30]. Perhaps these bacterial IMPDHs can catalyse both the IMPDH and GMPR reactions.

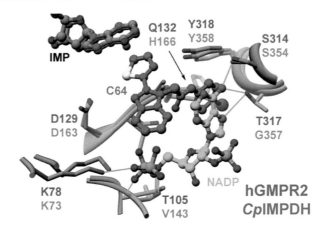

Figure 5. Cofactor binding sites in human GMPR2 and *C. parvum* IMPDH. The nicotinamide riboside portion of NADPH is omitted for clarity, as are the residues that interact with the 2'-phosphate (2c6q). *C. parvum* IMPDH inhibitor C 64 is shown (3khj). IMP (sticks) and inhibitor C 64 (ball and stick) are shown.

How does GMPR Distinguish Water and Ammonia?

The GMPR reaction involves the same E-XMP* intermediate as the IMPDH reaction, yet this intermediate does not react with water. In the presence of NADP+, E-XMP* forms readily from both IMP and 2-Cl-IMP [26]. XMP is more stable than GMP, so XMP would form if GMPR catalysed the reaction with water, but E-XMP* is stable for days in the absence of ammonia. Thus GMPR prefers ammonia over water by a factor of at least 10^6. What are the kinetic barriers that prevent GMPR from reacting with water?

The preference of GMPR for ammonia must derive from a combination of intrinsic reactivity and selective binding. Ammonia is a stronger nucleophile than water, by ~10^4–fold [31], but an additional factor of 100 is still required to account for the selectivity of GMPR. Ammonia has one more hydrogen bond donor than water, which could provide ~3 kcal/mol of specific binding energy, which could translate into a factor of 500 in selectivity. The ammonia channels AmtB and RhAG display ammonia/water selectivity of this order [32].

Was the Ancestral Enzyme a GMPR?

The primordial environment is believed to have been ammonia-rich and reducing [33], conditions that would favour the emergence of a GMPR operating to produce GMP. This hypothesis is supported by the observation that the over-expression of *E. coli* GMPR can complement bacteria lacking IMPDH and attenuated in GMPS [26]. Curiously, *Buchnera*, the aphid symbiont, is missing genes for both IMPDH and GMPS but contains a GMPR [34]. Insect guts are having high concentrations of ammonia/ammonium, reaching as high as

130 mM [35], so it is quite possible that GMPR is responsible for the synthesis of guanine nucleotides. Approximately 20 bacteria/archaea appear to be missing genes encoding GMPS, suggesting that these organisms also synthesize GMP directly from IMP and ammonia (The SEED, http://theseed.uchicago.edu/FIG/index.cgi accessed June 11, 2011).

If the ancestral enzyme was indeed a GMPR, then IMPDH would have evolved as the environment became oxidative and ammonia became limiting. Water activation would have initially depended on the Thr pathway, and therefore have involved a cofactor conformational change. However, with the installation of the Arg pathway for water activation, cofactor conformational changes would no longer be required. The hydride transfer step could be optimized and the cofactor binding site could migrate. Since Thr321 pathway is no longer required to activate water, the catalytic Glu431 could be substituted with Gln, which may represent a further specialization to the IMPDH reaction. We are now devising experiments to test the feasibility of this pathway.

ACKNOWLEDGEMENTS

This work was supported by NIH grant GM054403 (LH). Molecular graphics images were produced using the UCSF Chimera package from the Resource for Biocomputing, Visualization, and Informatics at the University of California, San Francisco (supported by NIH P41 RR001081).

REFERENCES

[1]   Glasner, M.E., Gerlt, J.A. and Babbitt, P.C. (2006) Evolution of enzyme superfamilies. *Curr. Opin. Chem. Biol.* **10**:492–497.
        doi: 10.1016/j.cbpa.2006.08.012.

[2]   Soskine, M. and Tawfik, D.S. (2010) Mutational effects and the evolution of new protein functions. *Nat. Rev. Genet.* **11**:572–582.
        doi: 10.1038/nrg2808.

[3]   Zalatan, J.G. and Herschlag, D. (2009) The far reaches of enzymology. *Nat Chem Biol.* **5**:516–520.
        doi: 10.1038/nchembio0809-516.

[4]   Nagano, N., Orengo, C.A. and Thornton, J.M. (2002) One fold with many functions: the evolutionary relationships between TIM barrel families based on their sequences, structures and functions. *J. Mol. Biol.* **321**:741–765.
        doi: 10.1016/S0022-2836(02)00649-6.

[5] Gerlt, J.A. and Raushel, F.M. (2003) Evolution of function in (beta/alpha)8-barrel enzymes. *Curr. Opin. Chem. Biol.* **7**:252 – 264.
doi: 10.1016/S1367-5931(03)00019-X.

[6] Wise, E.L. and Rayment, I. (2004) Understanding the importance of protein structure to nature's routes for divergent evolution in TIM barrel enzymes. *Acc. Chem. Res.* **37**:149 – 158.
doi: 10.1021/ar030250v.

[7] Lo Conte, L., Brenner, S.E., Hubbard, T.J., Chothia, C. and Murzin, A.G. (2002) SCOP database in 2002: refinements accommodate structural genomics. *Nucleic Acids Res.* **30**:264 – 267.
doi: 10.1093/nar/30.1.264.

[8] Orengo, C.A., Michie, A.D., Jones, S., Jones, D.T., Swindells, M.B. and Thornton, J.M. (1997) CATH – a hierarchic classification of protein domain structures. *Structure* **5**:1093 – 1108.
doi: 10.1016/S0969-2126(97)00260-8.

[9] Anantharaman, V., Aravind, L. and Koonin, E.V. (2003) Emergence of diverse biochemical activities in evolutionarily conserved structural scaffolds of proteins. *Curr. Opin. Chem Biol.* **7**:12 – 20.
doi: 10.1016/S1367-5931(02)00018-2.

[10] Hedstrom, L. (2009) IMP Dehydrogenase: structure, mechanism and inhibition. *Chem. Rev.* **109**:2903 – 2928.
doi: 10.1021/cr900021w.

[11] Ratcliffe, A.J. (2006) Inosine 5'-monophosphate dehydrogenase inhibitors for the treatment of autoimmune diseases. *Curr. Opin. Drug Discov. Devel.* **9**:595 – 605.

[12] Chen, L. and Pankiewicz, K.W. (2007) Recent development of IMP dehydrogenase inhibitors for the treatment of cancer. *Curr. Opin. Drug. Discov. Devel.* **10**:403 – 412.

[13] Olah, E., Kokeny, S., Papp, J., Bozsik, A. and Keszei, M. (2006) Modulation of cancer pathways by inhibitors of guanylate metabolism. *Adv. Enzyme Regul.* **46**:176 – 190.
doi: 10.1016/j.advenzreg.2006.01.002.

[14] Nair, V. and Shu, Q. (2007) Inosine monophosphate dehydrogenase as a probe in antiviral drug discovery. *Antivir. Chem. Chemother.* **18**:245 – 258.

[15] Hedstrom, L., Liechti, G., Goldberg, J.B. and Gollapalli, D.R. (2011) The antibiotic potential of prokaryotic IMP dehydrogenase inhibitors. *Current Medicinal Chemistry* **18**:1909 – 1918.
doi: 10.2174/092986711795590129.

[16]   Sun, X.E., Hansen, B.G. and Hedstrom, L. (2011) Kinetically controlled drug resistance: how *Penicillium brevicompactum* survives mycophenolic acid. *J. of Biol. Chem.* **286**:40595 – 40600.
doi: 10.1074/jbc.M111.305235.

[17]   Gan, L., Seyedsayamdost, M.R., Shuto, S., Matsuda, A., Petsko, G.A. and Hedstrom, L. (2003) The immunosuppressive agent mizoribine monophosphate forms a transition state analog complex with IMP dehydrogenase. *Biochemistry* **42**:857 – 863.
doi: 10.1021/bi0271401.

[18]   Kerr, K.M. and Hedstrom, L. (1997) The roles of conserved carboxylate residues in IMP dehydrogenase and identification of a transition state analog. *Biochemistry.* **36**:13365 – 13373.
doi: 10.1021/bi9714161.

[19]   Guillén Schlippe, Y.V. and Hedstrom, L. (2005) A twisted base? The role of arginine in enzyme-catalyzed proton abstractions. *Arch. Biochem. Biophys.* **433**:266 – 278.
doi: 10.1016/j.abb.2004.09.018.

[20]   Guillén Schlippe, Y.V. and Hedstrom, L. (2005) Is Arg418 the catalytic base required for the hydrolysis step of the IMP dehydrogenase reaction? *Biochemistry* **44**:11700 – 11707.
doi: 10.1021/bi048342v.

[21]   Guillén Schlippe, Y.V., Riera, T.V., Seyedsayamdost, M.R. and Hedstrom, L. (2004) Substitution of the conserved Arg-Tyr dyad selectively disrupts the hydrolysis phase of the IMP dehydrogenase reaction. *Biochemistry* **43**:4511 – 4521.
doi: 10.1021/bi035823q.

[22]   Guillén Schlippe, Y.V. and Hedstrom, L. (2005) Guanidine derivatives rescue the Arg418Ala mutation of *Tritrichomonas foetus* IMP dehydrogenase. *Biochemistry* **44**:16695 – 16700.
doi: 10.1021/bi051603w.

[23]   Min, D., Josephine, H.R., Li, H., Lakner, C., MacPherson, I.S., Naylor, G.J., Swofford, D., Hedstrom, L. and Yang, W. (2008) An enzymatic atavist revealed in dual pathways for water activation. *PLoS Biol.* **6**:e206.
doi: 10.1371/journal.pbio.0060206.

[24]   Spector, T., Jones, T.E. and Miller, R.L. (1979) Reaction mechanism and specificity of human GMP reductase. Substrates, inhibitors, activators, and inactivators. *J. Biol. Chem.* **254**:2308 – 2315.

[25] Deng, Y., Wang, Z., Ying, K., Gu, S., Ji, C., Huang, Y., Gu, X., Wang, Y., Xu, Y., Li, Y., Xie, Y. and Mao, Y. (2002) NADPH-dependent GMP reductase isoenzyme of human (GMPR2). Expression, purification, and kinetic properties. *Int. J. Biochem. Cell Biol.* **34**:1035 – 1050.
doi: 10.1016/S1357-2725(02)00024-9.

[26] Patton, G.C., Stenmark, P., Gollapalli, D.R., Sevastik, R., Kursula, P., Flodin, S., Schuler, H., Swales, C.T., Eklund, H., Himo, F., Nordlund, P. and Hedstrom, L. (2011) Cofactor mobility determines reaction outcome in the IMPDH/GMPR (β/α)8 barrel enzymes. *Nature Chem. Biol.* **7**:950 – 958.
doi: 10.1038/nchembio.693.

[27] Martinelli, L.K., Ducati, R.G., Rosado, L.A., Breda, A., Selbach, B.P., Santos, D.S. and Basso, L.A. (2011) Recombinant *Escherichia coli* GMP reductase: kinetic, catalytic and chemical mechanisms, and thermodynamics of enzyme-ligand binary complex formation. *Molecular BioSystems* **7**:1289 – 1305.
doi: 10.1039/c0mb00245c.

[28] Li, J., Wei, Z., Zheng, M., Gu, X., Deng, Y., Qiu, R., Chen, F., Ji, C., Gong, W., Xie, Y. and Mao, Y. (2006) Crystal structure of human guanosine monophosphate reductase 2 (GMPR2) in complex with GMP. *J. Mol. Biol.* **355**:980 – 988.
doi: 10.1016/j.jmb.2005.11.047.

[29] MacPherson, I.S., Kirubakaran, S., Gorla, S.K., Riera, T.V., D'Aquino, J.A., Zhang, M., Cuny, G.D. and Hedstrom, L. (2010) The structural basis of Cryptosporidium-specific IMP dehydrogenase inhibitor selectivity. *J. Am. Chem. Soc.* **132**:1230 – 1231.
doi: 10.1021/ja909947a.

[30] Gollapalli, D.R., Macpherson, I.S., Liechti, G., Gorla, S.K., Goldberg, J.B. and Hedstrom, L. (2010) Structural determinants of inhibitor selectivity in prokaryotic IMP dehydrogenases. *Chem. Biol.* **17**:1084 – 1091.
doi: 10.1016/j.chembiol.2010.07.014.

[31] Minegishi, S. and Mayr, H. (2003) How constant are Ritchie's "constant selectivity relationships"? A general reactivity scale for n-, pi-, and sigma-nucleophiles. *J. Am. Chem. Soc.* **125**:286 – 295.
doi: 10.1021/ja021010y.

[32] Musa-Aziz, R., Chen, L.M., Pelletier, M.F. and Boron, W.F. (2009) Relative CO_2/NH_3 selectivities of AQP1, AQP4, AQP5, AmtB, and RhAG. *Proc. Natl. Acad. Sci. U.S.A.* **106**:5406 – 5411.
doi: 10.1073/pnas.0813231106.

[33] Zahnle, K., Schaefer, L. and Fegley, B. (2010) Earth's earliest atmospheres. *Cold Spring Harb. Perspect. Biol.* **2**:a004895.
doi: 10.1101/cshperspect.a004895.

[34] van Ham, R.C., Kamerbeek, J., Palacios, C., Rausell, C., Abascal, F., Bastolla, U., Fernandez, J.M., Jimenez, L., Postigo, M., Silva, F.J., Tamames, J., Viguera, E., Latorre, A., Valencia, A., Moran, F. and Moya, A. (2003) Reductive genome evolution in *Buchnera aphidicola*. *Proc. Natl. Acad. Sci. U.S.A.* **100**:581–586.
doi: 10.1073/pnas.0235981100.

[35] Ji, R. and Brune, A. (2006) Nitrogen mineralization, ammonia accumulation, and emission of gaseous NH_3 by soil-feeding termites. *Biogeochemistry* **78**:267–283.
doi: 10.1007/s10533-005-4279-z.

[36] Gan, L., Petsko, G.A. and Hedstrom, L. (2002) Crystal structure of a ternary complex of *Tritrichomonas foetus* inosine 5'-monophosphate dehydrogenase: NAD^+ orients the active site loop for catalysis. *Biochemistry* **41**:13309–13317.
doi: 10.1021/bi0203785.

Beilstein-Institut

Experimental Standard Conditions of Enzyme Characterization
September 12th – 16th, 2011, Rüdesheim/Rhein, Germany

33

The pH-induced Metabolic Shift from Acidogenesis to Solventogenesis in *Clostridium acetobutylicum* – From Experiments to Models

Thomas Millat[1,*], Holger Janssen[2], Hubert Bahl[2], Ralf-Jörg Fischer[2] and Olaf Wolkenhauer[1,3]

[1]Dept. of Systems Biology & Bioinformatics, University of Rostock, 18051 Rostock, Germany

[2]Division of Microbiology, University of Rostock, 18051 Rostock, Germany

[3]Institute for Advanced Study (STIAS), Wallenberg Research Center, Stellenbosch University, Stellenbosch 7600, South Africa

E-Mail: *thomas.millat@uni-rostock.de

Received: 6th February 2012/Published: 15th February 2013

Abstract

The strictly anaerobic Gram-positive *Clostridium acetobutylicum* is able to ferment starchy material to acetone, butanol, and ethanol. Due to rising costs, dwindling resources, and environmental concerns regarding extraction and use of petroleum and natural gas, the academic and industrial interest in *C. acetobutylicum* has been renewed in recent years. However, an improved understanding of the clostridial metabolism and its regulations is a prerequisite for future industrial applications.

The COSMIC consortium, as part of the transnational SysMo initiative, is focusing on the pH-induced metabolic shift of *C. acetobutylicum* from acidogenesis to solventogenesis. During acidogenesis (high pH) the bacterium predominantly produces the acids acetate and butyrate, whereas the solvents acetone and butanol are fermented during solventogenesis (low pH). This metabolic phase transition is accompanied by changes in transcriptome, proteome, and metabolome which have been measured using a standardized experimental setup.

The information gathered is used to model this dynamic shift. Towards this end, we established a system of coupled differential equations, describing the biochemical reactions involved in AB fermentation and their dynamic changes found in recent experiments. Since *C. acetobutylicum* is not capable of maintaining a homoeostatic intracellular pH, the influence of a changing intracellular pH on enzyme activity and stability is of special interest for an improved understanding of AB fermentation. Such a model is able to predict product spectrum and metabolome during the pH-induced shift as well as for several mutants at solventogenesis.

INTRODUCTION

The obligatory anaerobic Gram-positive bacterium *Clostridium acetobutylicum* ferments starchy material to acetone, butanol, and ethanol which are important chemicals used in a plenty of industrial applications. The early industrial application if the acetone-butanol-ethanol (AB or ABE) fermentation is closely associated to Chaim Weizmann who helped to developed industrial facilities and held several patents related to this clostridial fermentation process. It became the second largest commercial biotechnological process ever performed until the early 1960 s when the production of these chemicals from petrochemicals became economically more favourable. Due to rising costs, dwindling resources, and environmental concerns regarding extraction and use of petroleum and natural gas, the academic and industrial interest in *C. acetobutylicum* has been renewed in recent years [1]. However, economic and ecological demands on the production of energy and (bio)chemicals require an optimization of the microbial fermentation. These demands include a large scale production of high-density, transportable energy that matches fossil fuels in convenience and utility. Here, *C. acetobutylicum* could be an adequate candidate for future CO_2-neutral energy production using (bio)butanol. First, it is non-pathogenic to humans, animals, and plants and produces no toxic by-products that may harm the environment. Second, butanol is a more attractive biofuel than ethanol which is currently used for that purpose [2, 3]. Its energy content is similar to that of pure gasoline and, thus, about 30% higher than that of ethanol. Furthermore, butanol is less corrosive, sparingly soluble in water, and does not absorb water. Hence, contrary to ethanol, butanol can be added at the refinery, be transported and delivered through existing infrastructure, and does not require modifications to engines. Despite the usage as biofuel, butanol is also an important bulk chemical needed for a broad range of applications including surface coatings, adhesives/scalings, elastomers, textiles, superabsorbents, flocculants, fibres, and plastics [4]. In summary, butanol is an important resource and clostridial AB fermentation could be an alternative to the use of dwindling petrol chemicals for its production.

However, the present state of technology is insufficient for a large-scale application of this bacterial process. Future usage requires an increased efficiency of the fermentation, for example the enhancement of metabolic flow through the pathway and butanol produced per glucose equivalent. Furthermore, the control of AB fermentation by external means is poorly understood. Moreover, the obligatory anaerobic feature of *C. acetobutylicum* and its preference to sugars as energy and carbon source hinder future applications. Hence, an improved understanding of clostridial metabolism and its regulations is a prerequisite for future large-scale industrial applications.

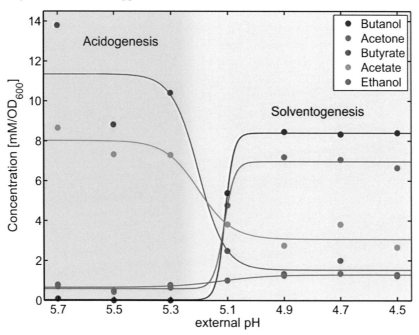

Figure 1. The steady-state product spectrum of AB fermentation in *C. acetobutylicum* as a function of the external pH. In a continuous culture under phosphate limitation, the bacterium changes its metabolic profile in response to variations of the external pH level. Experimental data (circles) are given in mM per optical density (OD_{600}). The subsequent normalization, using the optical density, was performed to account for fluctuations in the population size. Hyperbolic tangents are fitted to the experimental data (solid lines). Using their inflections points, we conclude that *C. acetobutylicum* is in acidogenic phase at pH < 5.1 and in solventogenic phase at pH < 5.1. A transition phase occurs in the range $5.2 <$ pH < 5.1.

Experimental evidence indicates that the extracellular pH level is crucial for the product spectrum and may, thus, enable an external control of the metabolic flow through the pathway. Interestingly, AB fermentation in *C. acetobutylicum* comprises two pH-dependent characteristic phases: acidogenesis and solventogenesis. During acidogenesis (high pH) the bacterium predominantly produces the acids acetate and butyrate, whereas the solvents acetone and butanol are fermented during solventogenesis (low pH). A systemic study of the steady-state product spectrum as a function of the external pH level is shown in Figure 1. Furthermore, we fitted hyperbolic tangents on the experimental data to define quantitative measures for this pH-dependent phase transition. Using their inflection points, we found that the bacterium is in acidogenic phase at pH > 5.2 and in solventogenic phase at pH < 5.1. An intermediary transition phase emerges within the range 5.2 < pH < 5.1 that differs from acidogenesis or solventogenesis. Furthermore, this metabolic phase transition is accompanied by changes in transcriptome, proteome, metabolome which have been measured using a standardized experimental setup.

The COSMIC consortium, as part of the European transnational SysMo initiative [5], is focusing on the pH-induced metabolic shift of *C. acetobutylicum* from acidogenesis to solventogenesis. A close collaboration between experimentalists and modellers realized an iterative cycle of data-driven models and model-driven design of experiments. This objective-driven approach will be discussed using the pH-induced metabolic shift in *C. acetobutylicum* as an example. Towards this end, we introduce recent and past experimental findings and the standardized experimental setup in the next section. Afterwards, we use this information to establish a dynamic mathematical model and discuss the assumptions applied in its formulation. Due to a lack of data, the pH-dependency of enzyme kinetic reactions involved in AB fermentation is neglected in current models. However, it is assumed that the pH-dependent specific activity is crucial for the kinetic regulation of the metabolic shift. Hence, we discuss its potential influence on the phase transition in the following section. Finally, we summarize our results and give an outlook to future research.

THE pH-INDUCED METABOLIC SHIFT AND THE METABOLIC NETWORK OF ABE FERMENTATION

In contrast to aerobic organisms that have established a prevalent tight regulation of intracellular pH [6–8], *C. acetobutylicum* is unable to maintain a constant intracellular pH, but rather the transmembrane pH gradient is kept constant [9, 10]. Experiments using batch and continuous cultures had been shown that the intracellular pH is higher with a difference $\Delta pH \approx 1$ [11, 12].

Under normal growth conditions *C. acetobutylicum* enters an exponential growth phase, which is characterized by the formation of acids, acetate and butyrate, as predominant liquid fermentation products. The increasing acid concentrations results in a decrease of the extracellular pH. Unable to maintain a homeostatic intracellular pH under this condition, the bacterium alters its metabolism to prevent a further reduction of the pH level. This phenomenon is referred to as the pH-induced metabolic shift and involves two processes: (i) the reutilization of previously secreted acids, and (ii) the production of pH-neutral solvents to prevent a recurrent decrease of the pH level. The data plotted in Figure 2 illustrate this phenomenon [11]. Furthermore, this batch experiment suggests that the internal pH follows the external one without significant delay.

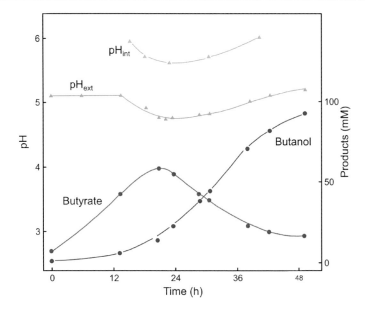

Figure 2. The extra- and intracellular pH and butyrate and butanol concentrations as a function of time in a growth experiment in batch culture. In contrast to aerobic organisms, the strict anaerobic *C. acetobutylicum* is unable to maintain a constant intracellular pH level, but rather keeps the transmembrane pH gradient constant. Thus, the intracellular pH follows the extracellular pH level without significant delay and an approximate difference ΔpH $\approx$ 1. Reproduced by courtesy of Springer from [11].

The preferred carbon and energy source of the bacterium is glucose, which is transported into the cell via a phosphoenolpyruvate-dependent phosphotransferase system [13, 14]. Then, the intracellular glucose is metabolized via glycolysis. The glycolytic steps form the backbone of acid and solvent formation and their activitiesare thus more or less pH-independent. Three key intermediates, acetyl-CoA, acetoacetyl-CoA, and butyryl-CoA, are of particular interest for AB fermentation [15]. These intermediates are branching points which channel the metabolic flow either to acid or solvent formation (Figure 3).

The first branching point, acetyl-CoA, is the starting point for the formation of acetate and ethanol. Acetate is generated in two sequential reactions involving a phosphotransacetylase and an acetate kinase [9, 16]. The activity of the acetate kinase rapidly decreases to very levels at solventogenesis [17, 18]. Simultaneously, ethanol is formed via an acetaldehyde/ alcohol dehydrogenase. Two antagonistically expressed NADH-dependent acetaldehyde/alcohol dehydrogenases, AdhE1 and AdhE2, play a crucial in the pH-induced metabolic shift [16, 19]. Whereas *adhe2* is transcribed during acidogenesis in a chemostat and its gene product only facilitates the formation of ethanol, AdhE1 is induced during solventogenesis and replaces AdhE2 [19]. Interestingly, the bifunctional AdhE1 promotes the production of both alcohols, ethanol and butanol [20].

The acetone formation starts from the second branching point, acetoacetyl-CoA, and is performed by the enzymes acetoacetyl-CoA transferase and acetoacetate decarboxylase, Adc, in two sequential reactions [9, 16]. The specific activity of Adc is very pH sensitive [21] and an increase by 38-fold from acidogenesis to solventogenesis was reported [17]. Furthermore, the uptake of acids during the shift requires active Adc [18, 22]. Interestingly, its concentration seems to more or less the same during both metabolic phases [23].

The last branching point, butyryl-CoA, is the starting point for the formation of either butyrate or butanol. Butyrate is produced in sequential reactions facilitated by two enzymes, phosphotransbutyrylase, Ptb, and butyrate kinase, which are most active during acidogenesis. Their specific activities decline during solventogenesis, 2-fold for Ptb and 6-fold for butyrate kinase [17]. The conversion of butyryl-CoA to butanol is promoted by sequential action of AdhE1. Importantly, experimental evidence indicates that AdhE1 is inactive in acid-producing cells, whereas its activity strongly increases during solventogenesis [19, 23, 24].

The involved CoA-transferase has fundamentally different role in clostridia compared to other bacteria [25]. Here, it is responsible for the uptake of formerly excreted acids, their conversion to the respective CoA-derivatives [17, 26], and acetone formation. In contrast to other enzymes involved in acid or solvent formation, it is insensitive to variations of the internal pH [27]. Because it operates in a sub-saturated regime, its activity is sensitive intracellular concentrations. Consequently, it is induced during solventogenesis.

The available information about ABE fermentation, summarized in the reduced network shown in Figure 3, suggests that the formation of acids and solvents is affected by pH-dependent kinetics and pH-dependent gene expression. Interestingly, the acid formation seems to be mostly modulated by changes of the kinetic properties, whereas the solvent formation requires both regulatory mechanisms.

Experimental evidence indicates that the pH-induced metabolic shift involves several levels of biological organization. However, the dynamic measurement of transcriptomic, proteomic, metabolic, and environmental changes is usually beyond the capacity of a single lab [28]. Thus, several labs with different specializations must closely collaborate to provide data fitting the needs of modelling. Towards this end, the COSMIC consortium established a standardized experimental setup and standard operation procedures.

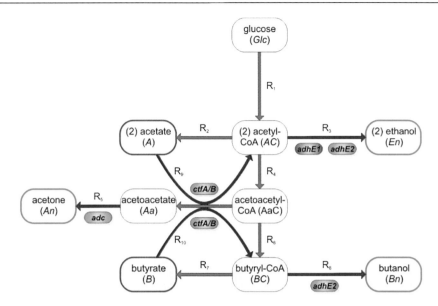

Figure 3. The reduced network of AB fermentation in *C. acetobutylicum*. The bacterium prefers glucose as carbon and energy source which is reduced in several glycolytic steps. During acidogenesis, the bacterium produces the acids acetate and butyrate (blue edges), whereas the solvents acetone and butanol (red edges) are the main liquid products during solventogenesis. This fermentative pathway comprises three key intermediate CoA-derivatives that are branching points of either acid or solvent production. Furthermore, recent experimental evidence indicates that several enzymes are induced in a pH-dependent manner. Those enzymes relevant for the AB fermentation are denoted on the reactions they facilitate. Furthermore, previously excreted acids are re-assimilated and converted to solvents during the metabolic shift. The abbreviations given in parentheses denote the variable names used in the mathematical model.

The continuous culture under phosphate limitation [29, 30] is superior to batch cultures, because it reduces the environmental and biological factors that could influence the bacterial population. Thus, it allows for a reasonable restriction of processes considered in the model and provides reproducible data which is a fundamental requirement for modelling. However, clostridial population may behave substantially different under these conditions than those in batch cultures [29]. The strain *C. acetobutylicum* ATCC 824 was grown under anaerobic conditions at $37\,^{\circ}\mathrm{C}$. The experiments were performed in a fermenter system, shown in Figure 4, with 0.5 mM KH_2PO_4 and 4% (wt/vol) glucose in the supplying medium and a dilution rate $D = 0.075\,\mathrm{h}^{-1}$. The external pH in the culture medium was adjusted to and kept constant at pH 5.7 (acidogenic phase) and pH 4.5 (solventogenic phase) by automatic addition of KOH.

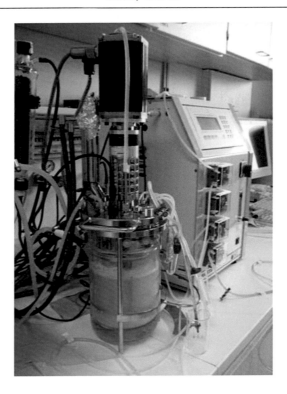

Figure 4. The standardized continuous culture setup used to investigate the influence of the external pH level on the metabolic state of *C. acetobutylicum*. The picture shows the well-stirred, pH-controlled fermenter and the control device which regulatesthe external pH. Photo by courtesy of Division of Microbiology, University of Rostock, Germany.

In three individual 'shift experiments', the pH level was changed from 5.7 to 4.5 to induce the metabolic shift. Data were taken for transcriptome, proteome, metabolome, product spectrum, and environome over the full length of the observation time. We used the gathered information to establish a mathematical model of the pH-induced metabolic phase transition in *C. acetobutylicum*. The model structure, underlying assumptions, and approximations are presented in the following section.

MODELLING THE DYNAMIC SHIFT FROM ACIDOGENESIS TO SOLVENTOGENESIS

Here we present a kinetic model of the AB fermentation in *C. acetobutylicum* in continuous culture. Because there is a lack of published information on the kinetic parameters governing these reactions under the conditions used in experiments in the literature, we aggregate a number of reactions of the metabolic network [9], see also Figure 3. This leads to a reduced number of model parameters that need to be estimated from experimental data. Simultaneously, we focus upon those reactions which are most likely to be regulated by the changing intracellular pH level. The resulting metabolic model is given by a coupled system of nine differential equations:

$$\frac{dAc}{dt} = R_1 - R_2 - R_3 - 2R_4 + R_9 - D \cdot AC,$$

$$\frac{dA}{dt} = R_2 - R_9 - D \cdot A,$$

$$\frac{dEn}{dt} = R_3 - D \cdot En,$$

$$\frac{dAaC}{dt} = R_4 - R_6 - R_9 - R_{10} - D \cdot AaC,$$

$$\frac{dAn}{dt} = R_5 - D \cdot An, \tag{1}$$

$$\frac{dBC}{dt} = R_6 - R_7 - R_8 + R_{10} - D \cdot BC,$$

$$\frac{dB}{dt} = R_7 - R_{10} - D \cdot B,$$

$$\frac{dBn}{dt} = R_8 - D \cdot Bn,$$

$$\frac{dAa}{dt} = R_9 + R_{10} - R_5 - D \cdot Aa.$$

Note that we rearranged the reactions in comparison to the representation we published in [31].

The rate of change of considered intracellular metabolites and products is determined by the sum of the reactions R_i that either produce (positive sign) or consume (negative sign) the respective molecule. In addition, we introduced an out-flow term which is the product of the dilution rate D with the concentration of the corresponding metabolite, because we have a constant out-flow of both extra- and intracellular, intracellular products as a result of cell out-flow through the fermenter.

Due to the aggregation of reactions, five glycolytic steps were combined into one reaction (R_i), adopting the assumption that there is a constant flux from glucose to acetyl-CoA. Additionally, we reduce the number of steps in five other reactions. Here, we assume that the corresponding intermediates are in quasi-steady state. Thus, in the conversions of two acetyl-CoA into either two molecules of acetate (R_2) or two molecules of ethanol (R_3), of acetoacetyl-CoA into acetone (R_5), and of butyryl-CoA into either butyrate (R_7) or butanol (R_8), we reduce two steps into one. Furthermore, we represent the three steps in the conversion of acetoacetyl-CoA to butyryl-CoA by one (R_6).

The resulting reaction rates considered in the metabolic network (1) are

$$R_1 = \frac{2V_1 \cdot Glc}{K_1 + Glc}, \qquad R_6 = \frac{V_6 \cdot AaC}{K_6 + AaC},$$

$$R_2 = \frac{V_2 \cdot AC}{K_2 + AC}, \qquad R_7 = \frac{V_7 \cdot BC}{K_7 + BC},$$

$$R_3 = a_3 \cdot AC \cdot adhe, \qquad R_8 = a_8 \cdot BC \cdot adhe, \qquad (2)$$

$$R_4 = \frac{V_4 \cdot AC}{2(K_4 + AC)}, \qquad R_9 = a_9 \cdot A \cdot AaC \cdot ctf,$$

$$R_5 = a_5 \cdot Aa \cdot adc, \qquad R_{10} = a_{10} \cdot B \cdot AaC \cdot ctf,$$

here we include the stoichiometric constant of two in R_1 since two molecules of acetyl-CoA are formed from one of glucose. Similarly, the constant 0.5 in R_4 represent the formation of one acetoacetyl-CoA from two acetyl-CoA. In consequence, we introduced a factor of two in the kinetic equation for acetyl-CoA, see metabolic network (1), to keep the correct stoichiometry for this molecule.

In Equations (2), Michaelis-Menten-like equations describe reactions involving enzymes with pH-independent gene regulation and, thus, with constant intracellular concentrations. This type of reaction is determined by an apparent limiting rate V_i and an apparent Michaelis-Menten constant K_i. In contrast, solving-producing enzyme are induced during the metabolic shift, so that the assumption of constant total enzyme concentrations made in the derivation of the Michaelis-Menten equation does not hold for the reactions those enzymes facilitate. Hence, we only apply the quasi-steady-state approximation which leads us to apparently bimolecular and trimolecular reactions, where the apparent rate coefficient is defined as $\alpha = k_3/K_m$ [31, 32].

In addition to the metabolic network (1), we require thus the following pH-dependent equations to describe the induction of solventogenic enzymes:

$$\frac{dadhe}{dt} = r_{\text{ah}} + r_{\text{ah}}^{+} \cdot F(pH) - D \cdot adhe,$$

$$\frac{dadc}{dt} = r_{\text{ad}} + r_{\text{ad}}^{+} \cdot F(pH) - D \cdot adc, \qquad (3)$$

$$\frac{dctf}{dt} = r_{ct} + r_{ct}^{+} \cdot F(pH) - D \cdot ctf.$$

These rate equations comprise two different synthesis rates. During acidogenesis, the enzymes are expressed with a low basal rate r_i. Their synthesis increases by r_i^{+} at solventogenesis. The pH-dependent function

$$F(pH) = -(pH(t) - 5.7)/1.2 \qquad (4)$$

generates the shift between both modes of gene expression. It is coupled on the external pH level which is approximated using a hyperbolic tangent [31].

Finally, we estimated all model parameters from three 'shift' experiments. Towards this end, we normalized and scale the data across the dynamic shift interval to make comparisons between time points meaningful. Each data set was interpolated at identical time points, enabling the average of the three scaled sets to be calculated for parameter estimation.

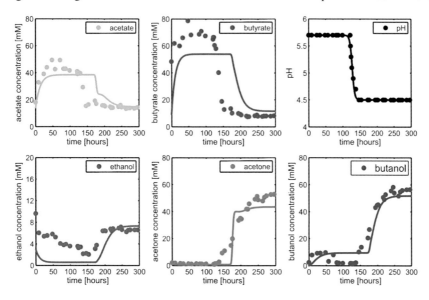

Figure 5. Comparison between experimental data and numerical simulation of a 'shift' experiment. The experiment (circles) started at acidogenic conditions (pH 5.7). During an initial phase (~100 hours), the culture established a steady state with the main products acetate and butyrate. At around 140 hours the pH control was switched off which results in the initiation of the metabolic switch. Finally, a solventogenic steady state is approached and the solvents butanol and acetone dominate the product spectrum. The results of the numerical simulation reproduce the measured steady states and describe the observed phase transition as a pH-induced rearrangement of the proteome involved in AB fermentation.

In Figure 5, we compare the numerical simulation results with data from a 'shift' experiment. The clostridial culture was maintained at acidogenesis (pH 5.7) for approximately 135 hours. During this period, an acidogenic steady state was established and the cells mainly produced the acids acetate and butyrate. Then, the pH control was stopped, allowing the natural pH-induced metabolic switch to solventogenesis to begin. During the shift, the product spectrum of the culture changed dramatically. The acid production dropped and the cells re-assimilated previously excreted acids from the medium resulting in a rapid decrease of acid concentration. Simultaneously, the formation of solvents was initiated and the concentrations of acetone and butanol increased strongly. This shift lasted about a day until the culture reached a solventogenic pH of 4.5. However, the continuous culture had not yet approached a metabolic steady state at this point. It took further 100 hours to establish a solventogenic steady state.

The steady-state product spectrum predicted by our numerical simulation is in accordance with the experimental data. This agreement proofs that the pH-induced shift involves a rearrangement of the proteomic composition of *C. acetobutylicum*. During the phase transition the bacteria induces additional solvent-producing enzymes. Interestingly, the concentration of acid-producing enzymes changes insignificantly [23], whereas the transcription is increased for several genes encoding those enzymes [24]. These findings may reflect the behaviour of *C. acetobutylicum* in its natural habitat. Here, the pH-induced metabolic shift is initiated to countervail against a further decline of the environmental pH level. Eventually, the bacterium returns to an acidogenic state that requires the presence of acid-producing enzymes. The found increase in transcription might, thus, compensate for decreased mRNA and or protein stability at solventogenesis. Furthermore, the simulation deviates from the experimental data during the dynamic transition from acidogenesis to solventogenesis. The theoretical predicted transition time is much shorter than measured in experiments. The pH-dependent kinetic enzymatic properties neglected in the present model could be a plausible explanation for that observation. In particular, the dynamics of acetone formation is overestimated which coincides with the fact that the involved acetoacetate decarboxylase exhibits a remarkable pH-dependent behaviour [17, 21]. This evidence suggests that the influence of the changing pH level on kinetic properties of the components involved in AB fermentation is crucial for the metabolic conversion processes as well as for their intracellular regulation. Thus, we investigate in the next section, how some important network motifs respond on variations in the intracellular pH if pH-dependent kinetics is considered. This will shed a new light on cellular processes related to environmental factors, like temperature, pH, or salinity that directly affect the enzymatic properties.

pH-DEPENDENT ENZYME KINETICS AND ITS EFFECT ON BRANCH POINTS OF THE ABE NETWORK

pH-dependence is not reflected in the common representation of the enzyme kinetic reaction shown in black in Figure 6, where the conversion of substrate S to product P is facilitated by enzyme E During this process an intermediary complex C is formed [33, 34]. The reaction becomes pH-dependent, if one considers that association and dissociation of hydrons, denoted in blue in Figure 6, change the structure of the enzyme and, thus, its specific activity.

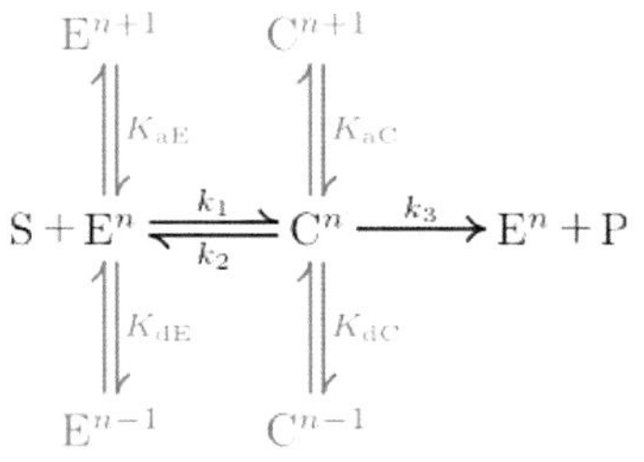

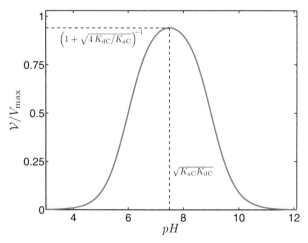

$$S + E^n \underset{k_2}{\overset{k_1}{\rightleftharpoons}} C^n \xrightarrow{k_3} E^n + P$$

Figure 6. The effect of the hydrons on the enzyme kinetic reaction. Due to association and dissociation of hydron to and from the enzyme and the enzyme-substrate complex the enzymatic activity is changed. Here, we assume that only enzymes with n bound hydrons are active and that association and dissociation of hydrons is much faster than the enzymatic conversion.

The incorporation of the pH-dependent association/dissociation of hydrons into the enzyme kinetic reaction leads to a formal equivalent expression for the reaction rate [34, 35]

$$\frac{dP}{dt} = -\frac{dS}{dt} = \frac{V_{\max} \cdot S}{K_m + S}, \tag{5}$$

with a pH-dependent limiting rate $V_{\max}$ and a pH-dependent Michaelis-Menten constant K_m. Both parameters are determined by the equilibrium of dissociation and association of hydrons which is described by the dissociation constants $K_{(a,d)E}$ and $K_{(a,d)C}$, where the subscript 'a' denotes the association of a hydron and the subscript 'd' the dissociation, and the hydron concentration H^+. The pH-dependent limiting rate [35]

$$V_{\max} = \frac{V_{\max}}{1 + H^+/K_{aC} + K_{dC}/H^+} \tag{6}$$

fulfils the relation $V_{\max} \leq V_{\max}$, where $V_{\max} = k_3 \cdot E^T$ is the limiting rate of the standard Michaelis-Menten equation. It exhibits a typical bell-shaped form, Figure 7.

Figure 7. The apparent limiting rate ν as a function of the pH exhibits a typical bell-shaped form. Its maximum is determined by the dissociation constants K_{aC} and K_{dC} that describe the association and dissociation of a hydron to and from, respectively, the complex C.

The pH-dependent apparent Michaelis-Menten constant [35]

$$K_m = K_m \frac{1 + H^+ / K_{aE} + K_{dE} / H^+}{1 + H^+ / K_{aC} + K_{dC} / H^+}. \tag{7}$$

results from the multiplication of the standard Michaelis-Menten constant $K_m = (k_2 + k_3)/k_1$ with a rational pH-dependent expression. A further analysis of rate equation (5) reveals that the pH-dependency of those apparent Michaelis-Menten constant has to be considered for small substrate concentrations $S \ll K_m$. Thus, we neglect its effect on the reaction rate in our further investigations. Then, the reaction rate is directly proportional to the pH-dependent limiting rate and shares the same functionality.

We now apply this formalism developed for isolated enzymes to investigate the effects of a changing pH on chosen network motifs of the AB fermentation pathway. Towards this end, we consider the separate impact of the intracellular pH on the regulation of gene expression using a (de)activation cycle with two distinct pH-dependent states, Figure 8[a], and on metabolic reactions using a branching point, Figure 8[b], as an example.

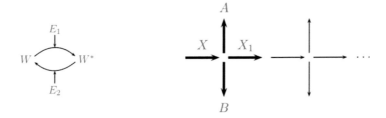

Figure 8. Schematic representation of two important motifs in the network of AB fermentation. Left: The (de)activation cycle of a protein W that comprises two distinct pH-dependent states. Here, we assume that the active form W^* may trigger the cellular adaption in response to changing pH levels, e.g. the induction of solvent-producing enzymes. The enzymes E_1 and E_2 activation and deactivate, respectively, the protein in pH-dependent reactions. Right: A metabolic branching point in the AB fermentation pathway. The intermediate X is converted either to product A or B in pH-dependent reactions promoted by the enzymes E_1 and E_2.

In the considered (de)activation cycle, a protein is activated by an enzyme E_1, e.g. by phosphorylation, and deactivated by enzyme E_2, e.g. by dephosphorylation, in a pH-dependent manner. Dependent on its activation state, the protein may trigger or inhibit specific cellular functions like transcription factor activity. Assuming that activation and deactivation are much faster than the cellular functions regulated by the protein, we focus on the steady state as a function of the external pH level and investigate how a shift of the position of the pH-dependent activity affects the steady-state concentration. For this purpose, we choose pH-dependent profiles with the same half width and height for activation and deactivation, respectively.

The steady-state concentrations as a function of the external pH are shown in Figure 9 for two different scenarios: [a] slightly differing pH-profiles, and [b] strongly differing pH-profiles. I accordance with experimental evidence, the latter setting assumes that the acid- and solvent-producing enzymes optimally operate either during acidogenesis or solvento-genesis. Because activation and deactivation processes possess different pH-dependent activity, the steady state is a function of the external pH. Due to the slightly differing pH-dependent activities investigated in Figure 9[a], the system 'fine-tunes' the protein concentration to its current requirements. In our example, the maximum of the activation is located at a smaller pH value than the deactivation. Thus, the active form of the protein is decreasing with increasing pH, whereas the inactive form increases with pH. Very different maxima result in pH-sensitive steady states that provide a switch-like behaviour, Figure 9[b]. The protein may, thus, act as an intracellular pH sensor which triggers metabolic and physiological adaptations in *C. acetobutylicum*.

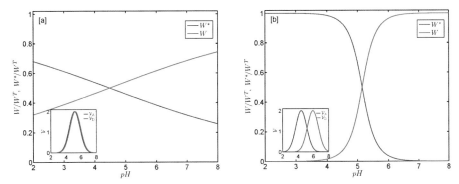

Figure 9. The fraction of the active and inactive form of protein W at steady state as function of the external pH. Two different situations are compared: **[a]** the maxima of the activities differ slightly ($M_A = 5.15$ and $M_D = 5.25$) and **[b]** the maxima differ-strongly from each other ($M_A = 5.45$ and $M_D = 5.95$). The corresponding specific activities, as a function of the pH, are shown as insets. With an increasing separation of the pH-profiles switch-like behavior emerges, which enables the (de)activation cycle to sense the intra- or extracellular pH level resulting an appropriate cellular response.

To investigate the influence of the intracellular pH on the product spectrum of a branching point, we consider a metabolite X, e.g. butyryl-CoA, which is converted to products A and B, e.g. butyrate and butanol. Additionally, metabolite X_1 shall be produced in both reactions that may be a reactant in further steps of a metabolic pathway, Figure 8[b]. The reactions are facilitated by enzymes E_1 and E_2. For the sake of simplicity, we assume that their concentrations are kept constant. Again, we consider two different scenarios, [a] slightly different specific activities and [b] strongly differing specific activities. Slightly differing specific result in a pH-dependent adaptation of the concentration of products A and B.

Interestingly, the concentration profile shows a switch-like behaviour for strongly differing activities. With an increasing pH level, the product which is produced in the enzyme kinetic reaction with the higher activity becomes dominant. In our example, this is the product B.

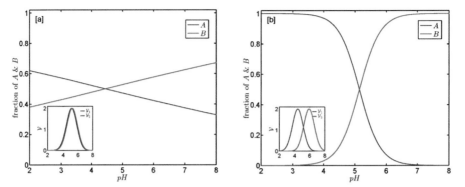

Figure 10. The fraction of products A and B with respect to the metabolite X as a function of the external pH value. Again, two situations are compared: **[a]** the maxima of the specific enzyme activities differ slightly and **[b]** the maxima differ strongly from each other. The corresponding specific activities as a function of the pH are shown in the insets. The diverging pH-dependent profiles lead to a pH-dependent product spectrum. Eventually, a changing pH shifts the product spectrum frommetabolite A to metabolite B and *vice versa*.

Interestingly, both scenarios representing genetic regulation and kinetic regulation exhibit a similar pH-dependent behaviour. However, they differ in their specific time scales. Kinetically regulated processes immediately respond to changes in the intracellular pH. In contrast, genetically regulated processes follow environmental variations with a delay caused by the time required for protein synthesis.

In summary, our investigation demonstrates that a changing pH level may lead to an altered product formation due to either changes in gene expression patterns or changes in specific activities of the reactions involved in the AB fermentation pathway of *C. acetobutylicum*.

Discussion

The bacterium *C. acetobutylicum* switches its metabolism between acidogenesis and solventogenesis in response to changes in the external pH. During acidogenesis (high pH), it dominantly produces acids, whereas the pH-neutral solvents acetone and butanol are the major fermentation products at solventogenesis. Several experimental studies, using a continuous culture under phosphate limitation, indicate that this metabolic adaptation involves changes on transcriptomic, proteomic, and metabolomic levels. However, the mechanisms underlying the regulation of the solventogenic shift remain unclear.

Using existent knowledge about the metabolic network and biochemical reactions, we established a kinetic model of AB fermentation in *C. acetobutylicum*. Incorporating gene regulation of solvent-producing enzymes furnishes a mechanistic representation of this pH-induced phase transition. Although, the model captures well the observed behaviour, including additional regulation may improve the fit. For instance, we have assumed that the induction of solventogenic enzymes is triggered at the same threshold pH value; assigning distinct threshold values to each encoding gene may improve the accuracy of the model.

Furthermore, considering that *C. acetobutylicum* is unable to maintain a constant intracellular pH level and experimental evidence indicates that the intracellular pH differs from the external pH by approximately $\Delta pH \approx 1$, we conclude that changes of the external pH affect the AB fermentation. Additionally, this idea is supported by observed strong pH-dependence of several enzymes involved in this metabolic process. Interestingly, the specific activity of the enzymes is optimal either during acidogenesis or solventogenesis. These findings suggest that pH-dependent kinetic properties of those enzymes may be important and could trigger the pH-induced metabolic switch.

Our investigations reveal that both regulatory mechanisms, genetic regulation, considered in the present model, and kinetic regulation may result in similar pH-dependent behaviour. Taken together experimental and theoretical knowledge, we conclude that kinetic and genetic regulation contribute to the cellular adaptation of *C. acetobutylicum* in response to changing pH levels.

Systems biology approaches involving alterations to gene expression levels are frequently adopted to investigate and exploit bacteria. For instance, we performed steady-state analyses to make predictions upon the effect of single mutations on the butanol yield during solventogenesis. By varying gene expression levels *in silico*, we infer that alterations in the expression of a single gene are insufficient to increase the butanol production significantly. Obviously, a more complex approach targeting two or more genes simultaneously is required.

The pH-dependency of enzymatic reactions has been neglected in previous models of clostridial AB fermentation potentially caused by missing knowledge about the pH-dependent kinetics of the enzymes and gene transcription. However, an isolated and independent consideration of kinetic and genetic regulation may be misleading. Further experimental and theoretical studies, therefore, require reliable information about pH-dependent enzymatic properties and pH-induced changes on transcriptomic, proteomic, and metabolomic levels. This information could provide the basis for a purposeful optimisation of the bacterium *C. acetobutylicum* for future industrial applications.

ACKNOWLEDGEMENTS

The authors acknowledge support by the German Federal Ministry for Education and Research (BMBF) as part of the European Transnational Network – Systems Biology of Microorganisms (SysMo) – within the COSMIC consortium (FKZ 0313981D and 0315782D). The responsibility for the content of this manuscript lies with the authors.

REFERENCES

[1]   Rittmann, B.E.(2008) Opportunities for renewable bioenergy using microorganisms. *Biotechnol. Bioeng.* **100**:203 – 212.
      doi: 10.1002/bit.21875.

[2]   Dürre, P. (2007) Biobutanol: An attractive biofuel. *Biotechnol. J.* **2**:1525 – 1534.
      doi: 10.1002/biot.200700168.

[3]   Green, E.M. (2011) Fermentative production of butanol – the industrial perspective. *Curr. Opin. Biotechnol.* **22**:337 – 343.
      doi: 10.1016/j.copbio.2011.02.004.

[4]   Dürre, P. (2008) Fermentative Butanol Production. *Ann. N.Y. Acad. Sci.* **1125**:353 – 362.
      doi: 10.1196/annals.1419.009.

[5]   http://www.sysmo.net

[6]   Booth, I.R. (1985) Regulation of Cytoplasmic pH in Bacteria. *Microbiol. Rev.* **49**:359 – 378.

[7]   Garcia-Moreno, B. (2009) Adaptations of proteins to cellular and subcellular pH. *J. Biol.* **8**:98.
      doi: 10.1186/jbiol199.

[8]   Krulwich, T.A., Sachs, G., and Padan, E. (2011) Molecular aspects of bacterial pH sensing and homeostasis. *Nat. Rev. Microbiol.* **9**:330 – 343.
      doi: 10.1038/nrmicro2549.

[9]   Jones, D.T. and Woods, D.R. (1986) Acetone-butanol fermentation revisited. *Microbiol. Rev.* **50**:484 – 524.

[10] Terracciano, J.S. and Kashket, E.R. (1986) Intracellular Conditions Required for Initiation of Solvent Production by *Clostridium acetobutylicum*. *Appl. Environ. Microbiol.* **52**:86 – 91.

[11]   Gottwald, M. and Gottschalk, G. (1985) The internal pH of *Clostridium acetobuty-licum* and its effect on the shift from acid to solvent formation. *Arch. Microbiol.* **143**:42–46.
doi: 10.1007/BF00414766.

[12]   Huang, L., Gibbins, L.N., and Forsberg, C.W. (1985) Transmembrane pH gradient and membrane potential in *Clostridium acetobutylicum* during growth under aceto-genic and solventogenic conditions. *Appl. Environ. Microbiol.* **50**:1043–1047.

[13]   Mitchell, W.J. (1997) Physiology of Carbohydrate to Solvent Conversion by Clos-tridia. *Adv. Microb. Physiol.* **39**:31–130.
doi: 10.1016/S0065-2911(08)60015-6.

[14]   Tangney, M. and Mitchell, W. (2007) Characterisation of a glucose phosphotransfer-ase system in *Clostridium acetobutylicum* ATCC 824. *Appl. Microbiol. Biotechnol.* **74**:398–405.
doi: 10.1007/s00253-006-0679-9.

[15]   Chen, J.-S. (1995) Alcohol dehydrogenase: multiplicity and relatedness in the sol-vent-producing clostridia. *FEMS Microbiol. Rev.* **17**:263–273.
doi: 10.1111/j.1574-6976.1995.tb00210.x.

[16]   Dürre, P., Fischer, R.-J., Kuhn, A., Lorenz, K., Schreiber, W., Stürzenhofecker, B., Ullmann, S., Winzer, K., and Sauer, U. (1995) Solventogenic enzymes of *Clostridium acetobutylicum*: catalytic properties, genetic organization, and transcriptional regula-tion. *FEMS Microbiol. Rev.* **17**:251–262.
doi: 10.1016/0168-6445(95)00006-X.

[17]   Andersch, W., Bahl, H., and Gottschalk, G. (1983) Level of enzymes involved in acetate, butyrate, acetone and butanol formation by *Clostridium acetobutylicum*. *Appl. Microbiol. Biotechnol.* **18**:327–332.
doi: 10.1007/BF00504740.

[18]   Hartmanis, M.G.N., Klason, T., and Gatenbeck, S. (1984) Uptake and activation of acetate and butyrate in *Clostridium acetobutylicum*. *Appl. Microbiol. Biotechnol.* **20**:66–71.
doi: 10.1007/BF00254648.

[19]   Fontaine, L., Meynial-Salles, I., Girbal, L., Yang, X., Croux, C., and Soucaille, P. (2002) Molecular Characterization and Transcriptional Analysis of adhE2, the Gene Encoding the NADH-Dependent Aldehyde/Alcohol Dehydrogenase Responsible for Butanol Production in Alcohologenic Cultures of *Clostridium acetobutylicum* ATCC 824. *J. Bacteriol.* **184**:821–830.
doi: 10.1128/JB.184.3.821-830.2002.

[20] Walter, K.A., Bennett, G.N., and Papoutsakis, E.T. (1992) Molecular characterization of two *Clostridium acetobutylicum* ATCC 824 butanol dehydrogenase isozyme genes. *J. Bacteriol.* **174**:7149–7158.

[21] Ho, M.-C., Menetret, J.-F., Tsuruta, H., and Allen, K.N. (2009) The origin of the electrostatic perturbation in acetoacetate decarboxylase. *Nature* **459**:393–397. doi: 10.1038/nature07938.

[22] Petersen, D.J. and Bennett, G.N. (1990) Purification of acetoacetate decarboxylase from *Clostridium acetobutylicum* ATCC 824 and cloning of the acetoacetate decarboxylase gene in *Escherichia coli*. *Appl. Environ. Microbiol.* **56**:3491–3498.

[23] Janssen, H., Döring, C., Ehrenreich, A., Voigt, B., Hecker, M., Bahl, H., and Fischer, R.-J. (2010) A Proteomic and Transcriptional View of Acidogenesis and Solventogenesis in *Clostridium acetobutylicum* in a Chemostat Culture. *Appl. Microbiol. Biotechnol.* **87**:2209–2226. doi: 10.1007/s00253-010-2741-x.

[24] Grimmler, C., Janssen, H., Krauße, D., Fischer, R.-J., Bahl, H., Dürre, P., Liebl, W., and Ehrenreich, A. (2011). Genome-Wide Gene Expression Analysis of the Switch between Acidogenesis and Solventogenesis in Continuous Cultures of *Clostridium acetobutylicum*. *J. Mol. Microbiol. Biotechnol.* **20**:1–15. doi: 10.1159/000320973.

[25] Wiesenborn, D.P., Rudolph, F.B., and Papoutsakis, E.T. (1989) Phosphotransbutyrylase from *Clostridium acetobutylicum* ATCC 824 and its role in acidogenesis. *Appl. Environ. Microbiol.* **555**:317–322.

[26] Hartmanis, M.G.N. and Gatenbeck, S. (1984) Intermediary Metabolism in Clostridium acetobutylicum: Levels of Enzymes Involved in the Formation of Acetate and Butyrate. *Appl. Environ. Microbiol.* **47**:1277–1283.

[27] Wiesenborn, D.P., Rudolph, F.B., and Papoutsakis, E.T. (1989) Coenzyme A transferase from *Clostridium acetobutylicum* ATCC 824 and its role in the uptake of acids. *Appl. Environ. Microbiol.* **55**:323–329.

[28] Williamson, W.P. (2005) Systems Biology: Will it Work? *Biochem. Soc. Trans.* **33**:503–506. doi: 10.1042/BST0330503.

[29] Fischer, R.-J., Oehmcke, S., Meyer, U., Mix, M., Schwarz, K., Fiedler, T., and Bahl, H. (2006) Transcription of the pst Operon of *Clostridium acetobutylicum* Is Dependent on Phosphate Concentration and pH. *J. Bacteriol.* **188**:5469–5478. doi: 10.1128/JB.00491-06.

[30]   Schwarz, K., Fiedler, T., Fischer, R.-J., and Bahl., H.A Standard Operating Procedure (SOP) for the preparation of intra- and extracellular proteins of *Clostridium aceto-butylicum* for proteome analysis. *J. Microbiol. Meth.* **68**:396 – 402. doi: 10.1016/j.mimet.2006.09.018.

[31]   Haus, S., Jabbari, S., Millat, T., Janssen, H., Fischer, R.-J., Bahl, H., King, J., and Wolkenhauer, O. A systems biology approach to investigate the effect of pH-induced gene regulation on solvent production by *Clostridium acetobutylicum* in continuous culture. *BMC Syst. Biol.* **5**:10. doi: 10.1186/1752-0509-5-10.

[32]   Millat, T., Bullinger, E., Rohwer, J., and Wolkenhauer, O. (2007) Approximations and their Consequences for Dynamic Modelling of Signal Transduction Pathways. *Math. Biosci.* **207**:40 – 57. doi: 10.1016/j.mbs.2006.08.012.

[33]   Cornish-Bowden, A. (2004). *Fundamentals of Enzyme Kinetics*.Portland Press.

[34]   Segel, I.H. (1993) *Enzyme Kinetics*. John Wiley and Sons.

[35]   Alberty, R.A. and Massey, V. (1954) On the interpretation of the pH variation of the maximum initial velocity of an enzyme-catalyzed reaction. *Biochim. Biophys.Acta* **13**:347 – 353. doi: 10.1016/0006-3002(54)90340-6.

SLOW-ONSET ENZYME INHIBITION AND INACTIVATION

ANTONIO BAICI

Department of Biochemistry, University of Zurich,
Winterthurerstrasse 190, CH-8057 Zurich, Switzerland

E-MAIL: abaici@bioc.uzh.ch

Received: 16th February 2012/Published: 15th February 2013

ABSTRACT

Interactions between modifiers and enzymes can either occur rapidly, on the time scale of diffusion-controlled reactions, or they can be slow processes observable on the steady-state time scale. Slow interactions in hysteretic enzymes serve to dampen cellular responses to rapid changes in metabolite concentration as part of regulatory mechanisms. Naturally occurring inhibitors of several enzymes, such as the macromolecular proteinaceous inhibitors of peptidases, may act slowly when forming complexes with their targets. To allow physiologically meaningful rates of enzyme inhibition, the modifier concentration is kept at high levels in nature but problems arise when these levels drop for some reason. The slow-onset inhibitory behavior of enzyme modifiers used as drugs may represent a handicap if their concentration at the target site is insufficient and/or the kinetic constants are inadequate to warrant pharmacologically meaningful rates of enzyme inhibition. A truthful knowledge of mechanisms and kinetic constants of such systems is mandatory for making predictions on the efficiency of the modifiers *in vivo*.

INTRODUCTION

Slow interactions in enzymology gained popularity after Frieden coined the term *hysteretic enzymes* for "... those enzymes which respond slowly (in terms of some kinetic characteristic) to a rapid change in ligand, either substrate or modifier, concentration" [1].

Frieden also derived an integrated rate equation, see equation (1) below, which relates the increase of product concentration with time. This equation was later shown to apply to a vast group of enzymatic mechanisms, whether or not enzyme hysteresis was involved, and the necessary mathematical background for inhibitors was further developed by Cha [2 – 4] and Morrison *et al.* [5 – 7].

Reversible enzyme inhibitors have been classified by Morrison in four categories (Table 1) depending on the rate of formation of their complexes with enzymes [5]. There are many instances in which the binding step of modifiers to enzymes occurs rapidly, whereas the sluggishness of the process is due to events other than the formation of the first complex. For this reason, the more general expression *slow-onset inhibition* will be used in place of slow-binding inhibition. The term 'slow' is vague as the time taken by these reactions varies from seconds to hours. However, it is generally agreed that slow-onset inhibition represents transient kinetics observed on the steady-state time scale and that the slowness of inhibition is understood in terms relative to the catalytic step, which is usually a faster process. There is no clear demarcation between the slow and fast categories in Table 1. For classical inhibitors the condition $[I] \approx [I]_t$ (the subscript means total concentration) could be trusted if it were not for the relative magnitudes of $[E]_t$, $[I]_t$ and K_i, which are not considered in Morrison's classification. In fact, a classical inhibitor can manifests tight-binding properties if $[E]_t$ is comparable in magnitude to K_i. On the other hand, typical high-affinity inhibitors show their efficiency already for $[I]_t$ $[E]_t$ at low enzyme concentrations to meet the condition $[I]_t$ $[E]_t$ K_i. Then, the relationships for the tight-binding and the slow, tight-binding classes of inhibitors in Table 1are better appreciated with the modification $[I]_t$ $[E]_t$ and K_i in place of $[I]_t$ $[E]_t$.

Table 1. Classification of the reversible enzyme inhibitors according to Morrison [5]. The original relationship between the total enzyme and inhibitor concentrations for the tight-binding and the slow, tight-binding cases, $[I]_t \approx [E]_t$, is changed here into $[I]_t \approx [E]_t$ and K_i.

Type of inhibition	Relationship between $[E]_t$ and $[I]_t$	Rate of formation of the inhibited complex
Classical	$[I]_t \gg [E]_t$	fast
Tight-binding	$[I]_t \approx [E]_t$ and K_i	fast
Slow-binding	$[I]_t \gg [E]_t$	slow
Slow, tight-binding	$[I]_t \approx [E]_t$ and K_i	slow

Irreversible enzyme inhibition, more appropriately called inactivation, is typically a slow process under *in vitro* conditions and can be treated analogously to slow-onset inhibition. The term *modifier* describes in general inhibition, inactivation and activation. This chapter is dedicated to slow-onset inhibition and inactivation, while slow-onset activation will not be treated. Steady-state and pre-steady-state equations for slow-onset activation have been derived by Hijazi and Laidler [8], and an overview of hysteretic, allosteric activators has been published by Neet *et al.* [9].

GENERAL ASPECTS OF SLOW-ONSET INHIBITION

There are several reasons why an inhibitor acts 'slowly', including (1) the existence of an intermediate whose structure recalls that of the transition state; (2) the inhibitor equilibrates rapidly between two or more forms, one of which makes up only a small proportion of all forms and interacts with the enzyme; (3) binding of the modifier to the enzyme to form an intermediate adsorption complex is a fast process, which is followed by a slow structural rearrangement to a second inhibitory complex; (4) only a rare form of the enzyme, which equilibrates between conformers, interacts with the modifier; (5) more trivially, it may be impossible to achieve sufficiently high modifier concentrations so that the rate of the second-order association reaction with enzyme is slow. The information supplied by the analysis of the transient phase in slow-onset inhibition experiments is superior to that gained from steady-state data because the exponential approach to steady-state can be used for calculating at least some individual rate constants for the inhibitory steps. The mechanism of inhibition and kinetic constants can be extracted from data either using integrated rate equations, when available, or by numerical integration.

For some mechanisms, integrated rate equations can be derived under restrictive assumptions but this is not always possible. The assumptions require $[S] \approx [S]_t$ and $[I] \approx [I]_t$, which means $[S]_t$, $[I]_t \gg [E]_t$, i.e. experiments must be properly designed to avoid excessive substrate turnover (say $\leq 10\%$) and to circumvent tight-binding between enzyme and inhibitor. In the presence of a slow-onset inhibitor, an enzyme-catalyzed reaction in which substrate is transformed into product (P) can be described by Frieden's equation mentioned above, which reads

$$[P] = v_s t + \frac{v_z - v_s}{\lambda} \left(1 - e^{-\lambda t}\right) + d, \tag{1}$$

where v_s and v_z are the velocities at steady-state and at time zero, respectively, and λ is the frequency constant of the exponential phase with reciprocal time as dimension. The parameter d, for displacement, was added to the original equation to account for any non-zero value of [P] or background of measured signal proportional to it at time zero. If an integrated rate equation exists and experiments can be set up to fulfill the assumptions, the equation can be fitted to progress curves by non-linear regression. In several cases, unambiguous diagnosis of the mechanism is possible by extracting the information contained in the expressions of the parameters v_s, v_z and λ. These are functions of inhibitor and substrate concentrations for a given mechanism and allow the calculation of rate constants. The graphical representation of slow-onset inhibition according to equation (1) is shown in Figure 1.

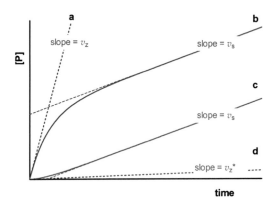

Figure 1. Progress curves for a generic slow-onset inhibition mechanism. The curve in (b) is obtained when substrate and inhibitor are mixed and reaction is started by adding enzyme. Curve (c) represents a reaction started by adding substrate to enzyme and inhibitor that had been preincubated for a sufficiently long tome to allow complex formation. The dashed line (a) is the tangent at time zero for curve (b) and dashed line (d) is the corresponding tangent for curve (c).

All mechanisms discussed below are drawn as the simplest one-substrate/one-product reaction for purely practical reasons with the implicit agreement that the same mechanisms can be applied to multi-reactant enzymes. For instance, in the case of a two-substrate/two-product enzymatic reaction, one substrate is kept at a constant concentration (not shown in the scheme) while the concentrations of the second substrate and inhibitor are varied. The measurements are then repeated exchanging the roles of the varied substrate.

The most frequent mechanism of slow-onset inhibition, both among naturally occurring and synthetic inhibitors, is of the linear competitive type. A general form is shown in Scheme 1, panel M1[1], in which no assumptions are made about the relative rates of equilibration of the two inhibitory steps. For this mechanism, a progress curve consisting of a double exponential followed by a linear steady-state release of product is expected but the derivation of an integrated rate equation is bound to restrictions, of which the absence of tight-binding between enzyme and inhibitor is the most important. However, it is right M1 in Scheme 1 that is often characterized by tight-binding.

---

[1] Reaction mechanisms, labeled Mx, where x is a progressive number, are grouped in this chapter as composite schemes to allow easier finding and cross-referencing

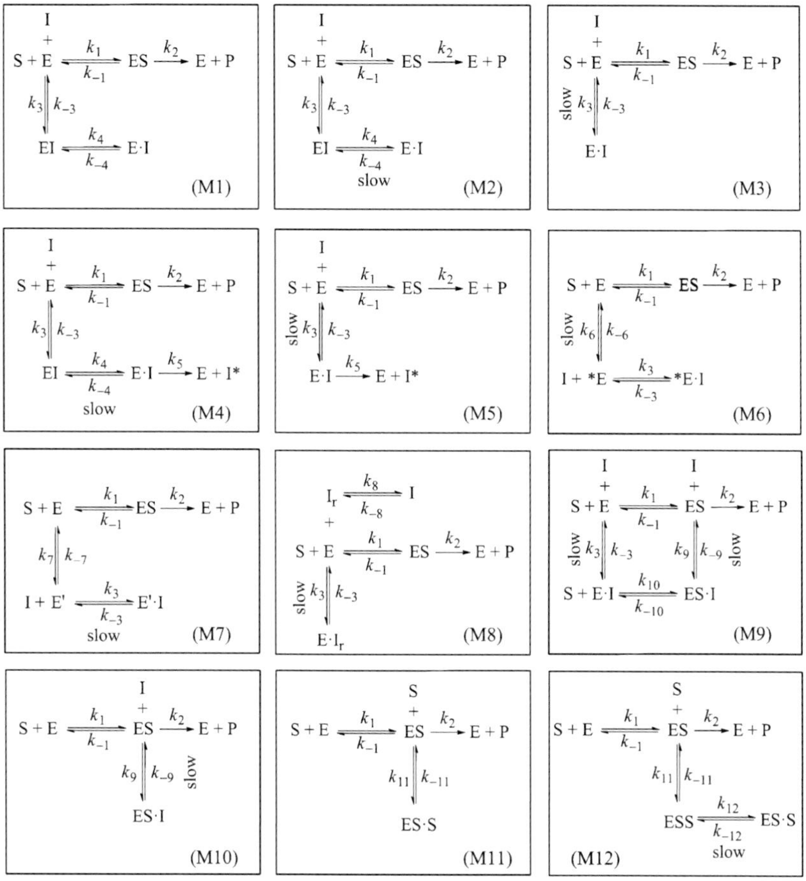

Scheme 1. Mechanisms giving rise to slow-onset inhibition. The labels M1 through M12 are introduced for facilitating cross-referencing, e. g. M3 is read 'mechanism 3'. E = enzyme, S = substrate, P = product, I = inhibitor. The steps labeled *slow* indicate qualitatively their relative rate of equilibration with respect to the other steps, which are assumed to be much faster. EI represents an adsorptive complex, while E•I, *E•I, E'•I, E•I$_r$, ES•I and ES•S denote reversible, non-covalent complexes. The numbers identifying kinetic constants of similar paths in diverse mechanisms are the same.

The first approach for analyzing the general mechanism M1 consists in fitting the generic equations (2) and (3) to progress curves and in evaluating which equation produces the best fit.

$$Y = A_1 \left(1 - e^{-\lambda_1 t} \right) + kt + d \tag{2}$$

$$Y = A_1 \left(1 - e^{-\lambda_1 t} \right) + A_2 \left(1 - e^{-\lambda_2 t} \right) + kt + d \tag{3}$$

Y is a signal proportional to product concentration, A_1 and A_2 represent amplitudes of the exponential phases, λ_1 and λ_2 are frequency constants, k is the slope of the straight line following the exponentials and d the value of Y at $t = 0$. If equation (3) fits data better than equation (2), any further use of equation (1) for non-linear regression analysis is discouraged. Instead, numerical integration is the method of choice in this case. However, even if the restrictions mentioned above can be avoided, this powerful approach may fail to extract the information from progress curves if λ_1 and λ_2 are *coupled*. That is to say, if both λ_1 and λ_2 contain the rate constants of the forward and reverse inhibitory reactions, the two steps of M1 cannot be separated from one another because they are 'mixed together'. Without analyzing in full the complexity of this system, such problems arise when the values of k_{-3} and k_4 are similar. Simulations of progress curves for this general slow-onset inhibition mechanism with various combinations of rate constants, reveals that the distinction between single exponential and double exponential reaction profiles is often very subtle and can be disclosed only after accurate statistical analysis. Fitting equations (1) or (2) generates progress curves that, by sight, appear nicely superimposed to data, though with worsening of the fit in dependence on inhibitor concentration if a single exponential is fit where a double exponential would better do the job. Introducing noise in the artificial data to simulate experimental error renders this distinction even more difficult and sometimes impossible. Yet, the most significant detail is that the kinetic constants calculated by fitting a single exponential equation to a biexponential progress curve can deviate considerably from the true values even if the fitted curve, as judged by inspection, appears to be almost perfect. It is difficult to ascertain from hundreds of reports in the literature whether the most appropriate model was fit to data because the overwhelming majority of published slow-onset inhibition cases were directly addressed to two variants of the general mechanism M1, namely M2 and M3 in Scheme 1, which can be analyzed with the monoexponential equation (1). These mechanisms are discussed in the next section.

A suggested approach to data analysis for the general mechanism M1 by numerical integration is shown in Figure 2. To appreciate the efficiency of the method, artificial data were simulated and random scatted was added to mimic experimental error. Two sets of fake data were produced, one for reactions started by adding enzyme (panel a) and another for reactions started with substrate (panel b). The idea is that data set (a) should represent more closely the association process, while data set (b) should give more information on the dissociation of the EI and E•I complexes. The set of differential equations for M1 was then globally fitted to all data by a combination of numerical integration and non-linear regression using KinTek software [10, 11]. When the kinetic constants k_3, k_{-3}, k_4 and k_{-4} were allowed to float freely during the global fit, it was impossible to obtain a unique solution; the values depended very much from the initial guesses and diverged considerably from the theoretical values used to generate the fake data. Data set (b) was then excluded from global fitting and only set (a) was analyzed in a first iteration until stable values of k_3 and k_{-3} were obtained. After constraining k_3 and k_{-3} to vary in a constant ratio, global fit was performed with data sets (a) and (b), which provided four kinetic constants very close to their theoretical values. Although the satisfactory overlap between data and best fits in Figure 2 is not yet a guarantee that the fits truly represent the system, running the FitSpace explorer [10] confirmed that all parameters were well constrained by data.

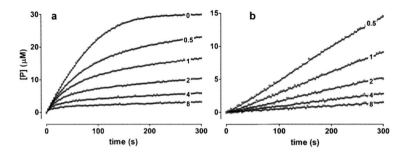

Figure 2. Analysis of progress curves for the general slow-onset inhibition mechanism (a) in Scheme 1. (a) Progress curves for reaction started by adding enzyme to a mixture of substrate and inhibitor. (b) Reactions started by adding substrate to enzyme and inhibitor preincubated for 300 s to allow complete equilibration of the EI and E•I complexes. The curves were simulated with the following constants and concentrations: $k_1 = 100\,\mu\mathrm{M}^{-1}\,\mathrm{s}^{-1}$, $k_{-1} = 1990\,\mathrm{s}^{-1}$, $k_2 = 10\,\mathrm{s}^{-1}$, $k_3 = 0.1\,\mu\mathrm{M}^{-1}\,\mathrm{s}^{-1}$, $k_{-3} = 0.05\,\mathrm{s}^{-1}$, $k_4 = 0.04\,\mathrm{s}^{-1}$, $k_{-4} = 0.005\,\mathrm{s}^{-1}$, $[\mathrm{E}]_t = 0.05\,\mu\mathrm{M}$, $[\mathrm{S}]_t = 30\,\mu\mathrm{M}$, $[\mathrm{I}]$ ($\mu\mathrm{M}$, the numbers shown in the two panels), $K_m = (k_{-1} + k_2)/k_1 = 20\,\mu\mathrm{M}$. Random scatter was added to simulate experimental error (red traces). The solid lines represent best fits obtained by combining numerical integration and non-linear regression using KinTek software [10, 11].

A critical inspection of Figure 2 shows that it had been impossible to analyze the progress curves with equation (1). In fact, the available substrate (30 μM) is used up to a great extent during reaction and, at least for the lowest inhibitor concentrations, a proportion of the inhibitor is bound to the enzyme, which cannot be neglected. The assumptions $[\mathrm{S}] \approx [\mathrm{S}]_t$ and $[\mathrm{I}] \approx [\mathrm{I}]_t$ are thus violated in at least some progress curves. As shown by trace '0' in panel (a), also the reaction with substrate alone can be included in the global fit, which gives further support to the quality of the analysis. The success in using curves with high substrate turnover for data analysis depends on product inhibition and this possibility can be considered in the model. It is also important to remark that in the progress curves containing two exponential phases, the fastest of these processes may be completed within a short time. Such experiments should be performed with a rapid-mixing device like the stopped-flow to minimize loss of essential data at the very beginning of the reaction. Particularly useful in this case is a logarithmic acquisition time to intensify data collection in the early phases of the reaction [12]. Using conventional photometry there is high risk to missing part or all of the first exponential, thus hampering proper analysis of the mechanism. Not shown here is another useful option, the inclusion in the global fit of progress curves obtained at fixed inhibitor concentration and variable substrate concentrations. This is useful to assess the character of the inhibition, i.e. for discriminating between competitive, mixed or uncompetitive inhibition. Published papers on 'linear competitive', slow-onset inhibition deserve as a remark that the competitive nature of inhibition was not always ascertained and analysis was performed by taking M2 or M3 in Scheme 1 for granted.

SPECIFIC CASES OF SLOW-ONSET INHIBITION

For all mechanisms discussed in this section, the integrated rate equation and the parameters therein presume the absence of inhibitor depletion after binding to the enzyme. This can experimentally be avoided if $[E]_t$ can be maintained at least one order of magnitude lower than $[I]_t$. Mechanisms M2 to M12 in Scheme 1 have in common the same integrated rate equation (1) but in some cases characteristic combinations of the expressions for v_s, v_z and λ. M2 and M3 can be distinguished from one another because λ depends linearly on [I] for M3 and hyperbolically for M2, while v_z depends hyperbolically on [I] in mechanism M2 while it is independent of [I] in M3 [equation groups (4) and (5)]. The kinetic constants for M2 can be calculated by non-linear regression analysis of λ versus [I], with k_{-4} representing the value of λ extrapolated for [I] = 0, and $(k_4 + k_{-4})$ as the asymptote of λ for [I]→∞. From the expression of v_z and v_s and the known value of [S] and K_m, K_i and the overall inhibition constant K_i^* can be extracted. Analogously for M3, k_{-3} is calculated as the intercept and k_3 from the slope of a plot of λ versus [I].

$$v_z = \frac{V[S]}{K_m\left(\frac{1+[I]}{K_i}\right)+[S]}; \qquad v_s = \frac{V[S]}{K_m\left(\frac{1+[I]}{K_i^*}\right)+[S]}; \qquad \text{(M2)}$$

$$\tag{4}$$

$$\lambda = k_{-4} + \frac{k_4[I]}{K_i\left(\frac{1+[S]}{K_m}\right)+[I]}; \quad K_i = \frac{k_{-3}}{k_3}; K_i^* = K_i\left(\frac{k_{-4}}{k_4+k_{-4}}\right)$$

$$v_z = v_0 = \frac{V[S]}{K_m+[S]}; \qquad v_s = \frac{V[S]}{K_m\left(1+\frac{[I]}{K_i}\right)+[S]}; \qquad \text{(M3)}$$

$$\tag{5}$$

$$\lambda = k_{-3} + \frac{k_3[I]}{1+\frac{[S]}{K_m}}; \qquad K_i = \frac{k_{-3}}{k_3}$$

M4 and M5 represent 'inhibitors' that are in reality substrates. At first sight they can be misidentified as true inhibitors because, for a given extent of time, progress curves are the same as those shown in Figure 1b. The expressions of v_s, v_z and λ are given in the equation groups (6) and (7), from which it is seen that discrimination between M2 versus M4 and M3 versus M5 is impossible using the dependence of v_z, v_s and λ upon [I] because the shapes of these plots are the same. However, if the reaction is started by adding substrate (curve c in

Figure 1), the slope of the linear part of the curve is independent of the preincubation time between enzyme and inhibitor for M2 and M3 while this increases with increasing preincubation time for M4 and M5. This because formation of I* reduces the amount of I that can react with the enzyme to form inhibited complex(es) thus favoring substrate turn-over. For integrating the rate equations of M4 and M5 the additional assumption $[I^*] \ll [I]_t$ is necessary, meaning that equation groups (6) and (7) are valid only if $[I]_t$ is sufficiently large and the measuring time sufficient short to satisfy this condition. The kinetic constants for M4 and M5 can be calculated likewise their respective counterparts M2 and M3, with the difference that only the sums $(k_{-4} + k_5)$ can be calculated for M4 and $(k_{-3} + k_5)$ for M5, not the individual values. These represent the net 'off' constants of the E●I complex. Examples of M4 and M5 can be found among polypeptides and proteins that bind to peptidases. Depending on the conditions, they can behave as inhibitors or can be cleaved as substrates as seen in the interaction between the thyroglobulin type-1 domain and the cysteine peptidase cathepsin L [13], which occurs according to mechanism M4. Although progress curves were shown without fitting to a particular equation, the slow-onset inhibitory interaction between Factor Xa and the leech-derived inhibitor antistasin appears to be a further example of mechanism M4 [14].

$$v_z = \frac{V[S]}{K_m\left(1+\dfrac{[I]}{K_i}\right)+[S]} \; ; \qquad v_s = \frac{V[S]}{K_m\left\{1+\dfrac{[I]}{K^*_{i,temp}}\right\}+[S]} \; ; \qquad \text{(M4)}$$

$$\text{(6)}$$

$$\lambda = k_{-4} + k_5 + \frac{k_4[I]}{K_i\left(1+\dfrac{[S]}{K_m}\right)+[I]} \; ; \quad K_i = \frac{k_{-3}}{k_3} \; ; \quad K^*_{i,temp} = K_i\frac{k_{-4}+k_5}{k_4+k_{-4}+k_5}$$

$$v_z = v_0 = \frac{V[S]}{K_m+[S]} \; ; \qquad v_s = \frac{V[S]}{K_m\left(1+\dfrac{[I]}{K^*_{i,temp}}\right)+[S]} \; ; \qquad \text{(M5)}$$

$$\text{(7)}$$

$$\lambda = k_{-3} + k_5 + \frac{k_3[I]}{1+\dfrac{[S]}{K_m}} \; ; \qquad K_{i,temp} = \frac{k_{-3}+k_5}{k_3}$$

The two particular cases of slow-onset inhibition M6 and M7 can be observed if an enzyme exists in equilibrium between different conformations and only one of these binds the inhibitor. The two mechanisms differ for the relative positions of the slow and fast steps. In M6 equilibration between two enzyme forms occurs slowly and only the state labeled *E reacts with the inhibitor. Although inhibitor binding may occur at the rate of diffusion, the overall inhibition process is slow because of the limited availability of the *E-form (the inhibitor 'must wait' for it). This mechanism it is likely to occur with enzymes that exist in distinct conformational states. M6 can be distinguished for the characteristic dependence of λ on $[I]$, which is a concave-up hyperbola (λ decreasing for increasing $[I]$) as shown in the equation group (8). While this allows the determination of k_6 and k_{-6}, it must be noted that M6 is a pretty complex mechanism because, depending on the equilibrium between E and *E, progress curves can show a lag both in the presence and in the absence of the inhibitor and v_z may or may not depend on [I] [6, 15].

In M7 the enzyme isomerizes in a fast process but inhibition proceeds slowly because the inhibitor can react only with a rare form E'. Since λ for M7 is a linear function of [I], the mechanism cannot be distinguished from M3 on the basis of this parameter, and also v_z and v_s are the same as for M3. In this case other sources of information, such as knowledge of enzyme structure from crystallography or different spectroscopic properties of E and E', must be invoked.

$$\lambda = \frac{k_6}{1+\frac{[S]}{K_m}} + \frac{k_{-6}}{1+\frac{[I]}{K_i}}; \quad K_i = \frac{k_{-3}}{k_3} \quad \text{(M6)} \tag{8}$$

$$\lambda = k_{-3}\frac{1+\frac{[S]}{K_m}+\frac{[I]}{K_i}}{1+\frac{[S]}{K_m}}; \quad v_z \text{ and } v_s \text{ as in M3} \quad \text{(M7)} \tag{9}$$

$$K_m = \frac{(k_{-1}+k_2)(k_7+k_{-7})}{k_1 k_7}; \quad K_i = \frac{k_{-3}(k_7+k_{-7})}{k_3 k_7}$$

M8 is analogous to M7 but in this case slow-onset inhibition is due to equilibration of the inhibitor between different molecular forms of which only a rare species reacts with the enzyme. A well-documented case of M8 is the inhibition of cathepsin B by the peptide aldehyde leupeptin [16, 17]. The slow-onset inhibitory behavior of this system, originally attributed to a hysteretic effect on the part of the enzyme [16], was later demonstrated by NMR to be due to equilibration of leupeptin in aqueous solution between three forms: a cyclic carbinolamine (42%), a leupeptin hydrate (56%) and the free aldehyde (2%) [17]. Only the free aldehyde behaves as inhibitor, which represents 2% of the mixture at any concentration. As shown in equation group (10), the dependence of λ on [I] for M8 is linear as it is for M3 in equation group (5). Hence, the two mechanisms cannot be distinguished from one another by kinetic measurements. The effective concentration of the inhibitor must be measured by an independent method that allows calculation of the equilibrium constant K_r. Since $K_r \ll 1$, the effective inhibitor concentration is reduced by the factor K_r, which explains slow-onset inhibition.

$$\lambda = k_{-3} + \frac{k_3 K_r [I]}{1+\frac{[S]}{K_m}}; \quad K_r = \frac{[I_r]}{[I]} = \frac{k_{-8}}{k_8} \ll 1 \quad \text{(M8)} \tag{10}$$

A neglected aspect is the possible occurrence of mixed inhibition (M9) besides the widely reported competitive inhibition. Equations for this system are shown in the equation group (11), which show that λ is a linear function of [I], v_z is independent of [I] and v_s is hyperbolically dependent on [I]. Although these patterns are indistinguishable from those of M3, the dependence of λ on [S] is different for M3 and M9. Thus, a set of progress curves at fixed inhibitor concentration and variable [S] will show the identity of the mechanisms. The systematic measurement of the substrate dependence of progress curves for slow-onset inhibition is anyhow highly recommended for both diagnostic and computational purposes.

$$v_z = \frac{V[S]}{K_m + [S]}; \qquad v_z = \frac{V \frac{[S]}{K_m}}{1 + \frac{[I]}{K_i} + \frac{[S]}{K_m}\left(1 + \frac{[I]}{\alpha K_i}\right)}; \quad \text{(M9)}$$

$$\tag{11}$$

$$\lambda = \frac{k_{-3} + \frac{k_{-9}k_{10}}{k_{-10}}[S]}{1 + \frac{k_{10}}{k_{-10}}[S]} + \frac{k_3 + k_9 \frac{[S]}{K_m}}{1 + \frac{[S]}{K_m}}[I]; \quad K_i = \frac{k_{-3}}{k_3}; \quad \alpha K_i = \frac{k_{-9}}{k_9}$$

M10 represents slow-onset uncompetitive inhibition, which was shown to occur in enzymatic reactions involving two substrates and two products. Inhibition of the enoylreductase FabI from *Escherichia coli* by triclosan belongs to this type [18] although, for technical reasons, the authors could not use progress curves for analyzing slow-onset inhibition. Equation group (12) lists the expressions that apply to M10.

$$v_z = \frac{V[S]}{K_m + [S]}; \quad v_s = \frac{V[S]}{K_m + [S]\left(1 + \frac{[I]}{\alpha K_i}\right)}; \quad \text{(M10)}$$

$$\tag{12}$$

$$\lambda = k_{-9} + \frac{k_9[I]}{1 + \frac{K_m}{[S]}}; \qquad \alpha K_i = \frac{k_{-9}}{k_9}$$

The dependence of λ on [I] is linear as it is for M3 but the dependence of λ on [S] can discriminate between the two cases because λ decreases or increases with increasing [S] for M3 or M10, respectively. These diagnostic properties and the way kinetic constants can be extracted are displayed in Figure 3.

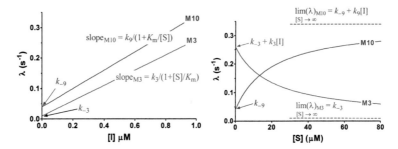

Figure 3. Dependence of the frequency constant λ on substrate and inhibitor concentration for mechanisms M3 and M10. The characteristic intersection points on the ordinate in both panels, as well as the asymptotes for $[S] \to \infty$ in the right panel (dashed lines) allow calculation of kinetic constants. Curves were simulated with the following parameters: $k_3 = 0.5\,\mu M^{-1}\,s^{-1}$, $k_{-3} = 0.01\,s^{-1}$, $k_9 = 0.6\,\mu M^{-1}\,s^{-1}$, $k_{-9} = 0.04\,s^{-1}$, $K_m = 20\,\mu M$. The fixed $[S]$ in the left panel was $20\,\mu M$, and the fixed $[I]$ in the right panel was $0.5\,\mu M$.

Inhibition by excess substrate, as shown in M11, is formally equivalent to uncompetitive inhibition. Classically, substrate inhibition has been treated as a fast process. However, it is possible that binding of substrate molecules to a site different from the catalytic center exhibit the properties of slow-onset inhibition, if nothing else because the secondary binding may have lesser affinity than that at the catalytic center. This means that the secondary binding may be slow because saturation of this process requires very high substrate concentrations and, while progressively increasing $[S]$ in an experiment, a threshold is reached for which substrate begins to appreciably bind to it but this concentration is still too low to allow the path $S + ES \to ES \bullet S$ to be sufficiently fast. M12 is a degenerated form of M11 in the same way as M3 is a degenerated form of M2. M12 was documented for the enzyme cathepsin K, which owns an exsosite to which ligands, including the substrate itself, can bind [19]. The expressions of the parameters in equation (1) for M11 and M12 are given in the equation groups (13) and (14), respectively.

$$v_z = \frac{V[S]}{K_m + [S]}; \qquad v_s = \frac{V[S]}{K_m + [S]\left(1 + \dfrac{[S]}{K_{si}}\right)} \qquad \text{(M11)}$$

$$\lambda = k_{-11} + \frac{k_{11}[S]}{1 + \dfrac{K_m}{[S]}}; \qquad K_{si} = \frac{k_{-11}}{k_{11}} \qquad (13)$$

$$v_z = \frac{V[S]}{K_m + [S]\left(1 + \dfrac{[S]}{K_{si}}\right)}; \qquad v_s = \frac{V[S]}{K_m + [S]\left(1 + \dfrac{[S]}{K_{si}^*}\right)} \qquad \text{(M12)}$$

$$\lambda = k_{-12} + \frac{k_{12}[S]}{K_{si}\left(1 + \dfrac{K_m}{[S]}\right) + [S]}; \qquad K_{si} = \frac{k_{-11}}{k_{11}}; \quad K_{si}^* = K_{si}\frac{k_{-12}}{k_{-12} + k_{12}} \qquad (14)$$

Other mechanisms for slow-onset inhibition besides those listed in Scheme 1 are not shown here. These are likely to occur in practice but not all of them can be unambiguously identified if not supported by adequate and sufficiently precise data.

The tight-binding condition was not treated in this section devoted to the use of analytical solutions of differential equations. Indeed, tight-binding represents just an experimental issue, not an intrinsic property of the mechanisms. Furthermore, an analytical, integrated rate equation that takes into account inhibitor depletion can be derived for M3 but not for M2 [3], while for the other mechanisms in Scheme 1 this point was not explicitly addressed in the literature. Although this derivation is possible for some systems, the necessary assumptions are too restrictive to be useful in practical situations. Alternatively, numerical integration methods can be used in place of non-linear regression. For this purpose, a system of dedicated differential equations can easily be written for each one of the mechanisms in Scheme 1 and any other mechanism of this type. This labor is even superfluous when using modern software [10, 11]. The issue is that several models must be applied until the best one is found but this may be a tedious and time-consuming approach. As far as the pre-steady-state of progress curves consist of a single exponential, i.e. as long as equation (2) fits data better than equation (3), the identification of the appropriate model for numerical integration is possible by preliminary analysis with the diagnostic criteria outlined above using analytical solutions. Analyzing the parameter dependence on [I] and [S] will give hints as to which mechanism comes closer to that described by the experiment, no matter if too much substrate is turned over during the observation time and if plotting parameters against $[I]_t$ instead of [I] will produce bias in the values of the kinetic constants. These can be used later as initial guesses for refinement by numerical integration once the appropriate model has been identified.

ENZYME INACTIVATION

Modifiers can react with elements of the catalytic center of enzymes forming covalent bonds that result in enzyme inactivation. Common mechanisms observed experimentally for this category of modifiers are listed in Scheme 2 (M13 – M18), where the symbol E-I distinguishes at a glance inactivation from reversible inhibition. Typically, E-I covalent bonds are formed slowly and reactions can require minutes to hours to proceed to completeness depending on the characteristic constants of the system and reactant concentrations. For M13 and M14 the integrated rate equation is (15) and the parameters are given in the equation groups (16) and (17):

$$[P] = \frac{v_z}{\lambda}\left(1 - e^{-\lambda t}\right) + d, \tag{15}$$

$$v_z = \frac{V[S]}{K_m\left(1+\dfrac{[I]}{K_i}\right)+[S]}; \quad \lambda = \frac{k_4[I]}{K_i\left(1+\dfrac{[S]}{K_m}\right)+[I]}; \quad K_i = \frac{k_{-3}}{k_3}. \quad \text{(M13)} \tag{16}$$

$$v_z = v_0 = \frac{V[S]}{K_m+[S]}; \quad \lambda = \frac{k_3[I]}{1+\dfrac{[S]}{K_m}} \quad \text{(M14)} \tag{17}$$

Equation (15) differs from (1) for the absence of the term v_s because k_{-4} and k_{-3} equal zero in M13 and M14, respectively. The diagnosis of these mechanisms is generally easy because a steady-state is absent in the progress curves (the single exponent levels off to a line parallel to the time axis for increasing time) and because plots of λ versus [I] intersect the ordinate at zero after any correction for the term d in equation (15). In general however, this last criterion should not be taken as a demonstration of irreversibility because small values of k_{-4} and k_{-3} in mechanisms M2, M3 and others, particularly in presence of experimental scatter, can render impossible the distinction of a small value of the ordinate intercept from zero.

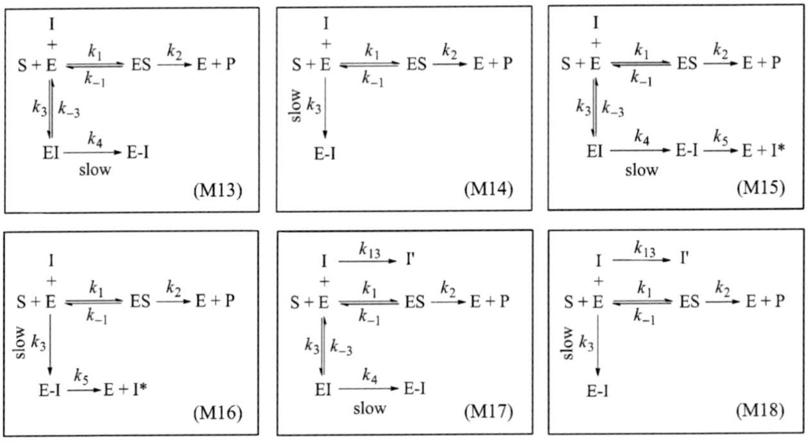

Scheme 2. Mechanisms of enzyme inactivation. Labeling as M13 – M18 continues the list of Scheme 1 and numbering of kinetic constants reproduces similar paths for the reversible counterparts in Scheme 1. E = enzyme, S = substrate, P = product, I = inactivator. E – I denotes a covalent bond between enzyme and inactivator.

Equation (15) is invalid for the temporary inactivation mechanisms M15 and M16 because the recycled free enzyme can again combine with substrate and inactivator. It is assumed that the enzymatically transformed inactivator, I*, has lost any affinity for the enzyme. The appropriate equation is (18), which contains the term v_∞ for the velocity 'at the end' of the exponential phase, i.e. $t = \infty$ in $e^{-\lambda t}$ [20]:

$$[P] = v_\infty t + \frac{v_z - v_\infty}{\lambda}\left(1 - e^{-\lambda t}\right) + d. \tag{18}$$

The velocity term v_∞ resembles v_s in equation (1) and progress curves have the same shape as in Figure 1b, which give the impression of a reversible mechanism. Without knowledge of the temporary character of the inactivation from chemical or other information, an impulsive diagnosis of such progress curves would suggest either M2 or M3 as candidate mechanisms. This is evidenced from the expressions of λ in equation groups (19) and (20), which are formally identical to (4) and (5), respectively.

$$v_z = \frac{V[S]}{K_m\left(1+\dfrac{[I]}{K_i}\right)+[S]}; \qquad v_\infty = \frac{V[S]}{K_m\left\{1+\dfrac{[I]}{K_i}\left(1+\dfrac{k_4}{k_5}\right)\right\}+[S]} \qquad \text{(M15)}$$

$$\qquad\qquad\qquad\qquad\qquad\qquad\qquad\qquad\qquad\qquad\qquad\qquad (19)$$

$$\lambda = k_5 + \frac{k_4[I]}{K_i\left(1+\dfrac{[S]}{K_m}\right)+[I]}; \qquad K_i = \frac{k_{-3}}{k_3}$$

$$v_z = v_0 = \frac{V[S]}{K_m+[S]}; \qquad v_\infty = \frac{V[S]}{K_m\left(1+\dfrac{k_3}{k_5}[I]\right)+[S]} \qquad \text{(M16)}$$

$$\qquad\qquad\qquad\qquad\qquad\qquad\qquad\qquad\qquad\qquad\qquad\qquad (20)$$

$$\lambda = k_5 + \frac{k_3[I]}{1+\dfrac{[S]}{K_m}}$$

Mechanisms M15 and M16 are different from mechanism-based (suicide) inhibition, which is not treated here. A way for discriminating kinetically M15/M16 from M2/M3 is to run experiments with enzyme and inactivator preincubated for various times, preferentially with low inactivator concentrations, and to start reactions by adding substrate. Provided the enzyme is stable during the measuring time, mechanisms M15 and M16 will show preincubation time-dependent regain of enzyme activity until reaching the rate v_0 after complete transformation of all added inactivator to the inert species I*. This cannot happen with mechanisms M2 and M3 but still the possibility of having to do with the reversible temporary mechanisms M4 and M5 cannot be ruled out, and here we have reached an end point in which enzyme kinetics must seek help from other methods for analyzing the fate of the inactivator after having been in contact with the enzyme. Examples are known from many published studies, such as a large screening of phosphadecalin derivatives as inactivators of acetylcholinesterase [20, 21], in which kinetic measurements were supported by NMR spectroscopy (see [20] and references therein).

The last two inactivation mechanisms in Scheme 2, M17 and M18, apply to unstable inactivators, to which a meticulous theoretical work has been dedicated [22]. The instability of some compounds in aqueous solution is often due to hydrolysis that leads to chemically transformed, inert species (I' in Scheme 2), which may represent a limitation for their practical use. Progress curves for M17 and M18 resemble those for reversible inhibition or temporary inactivation, i.e. they consist of an exponential burst followed by a linear increase of product release as depicted in Figure 1b. An integrated rate equation cannot be derived for these cases but Topham [22] provided the following analytical approximation for M18 based on Maclaurin expansion:

$$[P] = \frac{v_z}{k_{13}} e^{-A[I]_t/k_{13}} \left\{ k_{13}t + \sum_{i=1}^{\infty} \left(\frac{A[I]_t}{k_{13}}\right)^i \frac{\left[1-\left(e^{-k_{13}t}\right)^i\right]}{i \times i!} \right\} \quad (M18)$$

$$(21)$$

$$A = \frac{k_3 K_m}{K_m + [S]}$$

A useful property of this treatment is that equation (21) applies to competitive, uncompetitive and mixed inactivation by an unstable modifier. Moreover, the same equation is suitable for analyzing the effects of unstable activators. Every mechanism owns a dedicated expression of the apparent second-order rate constant A in equation (21) [22]. These and other mechanisms are not discussed in this chapter.

For M18, the rate constants k_3 and k_{13} can be calculated by non-linear regression fit of equation (21) to data with appropriate adjustment of the number of terms (i) in the Maclaurin expansion. The choice of the necessary terms depends on the value of k_{13}. For instance, with $k_{13} = 0.005 \text{ s}^{-1}$ the third term is sufficient, whereas with $k_{13} = 0.001 \text{ s}^{-1}$ expansion must be performed to the 10th term [20]. The complex analytical solution for M17, also provided by Topham [22], may overwhelm the endurance of end users less experienced in enzyme kinetics, in which case numerical integration can be managed easier.

THE PHYSIOLOGICAL AND PHARMACOLOGICAL SIGNIFICANCE OF SLOW-ONSET ENZYME-MODIFIER INTERACTIONS

Mechanisms M2 – M18 have in common a slow monoexponential phase for either inhibition or inactivation, for which the half-time can be calculated as $t_{1/2} = \ln 2 / \lambda$ [23], where λ is the frequency constant in equations (4) – (20). After seven half-times the exponential phase for enzyme-modifier association is more than 99% complete and a *delay time for inhibition/ inactivation* can be defined as DTI $= (7 \times 0.693)/\lambda$ or

$$DTI \approx 5/\lambda. \quad (22)$$

This is the time needed by both naturally occurring and exogenous modifiers used as drugs for neutralizing unwanted enzyme activities. Knowing the inhibition or inactivation mechanism, the related kinetic constants, the substrate concentration and K_m, λ can be calculated and then introduced in (22) for computing the DTI. From the kinetic parameters, the extent of substrate turnover during the DTI can be calculated and this information can be used to estimate the required modifier concentration at the target site for hindering unwanted effects in due time [24]. Even knowing an approximate value of DTI can be of great help because the successful use of enzyme modifiers as drugs depends on modifier bioavailability, concentration at the target site and mechanism of inhibition/inactivation. Specifically, if depends linearly on [I], DTI can be made as short as the modifier concentration can be increased because DTI$\rightarrow$ 0 for [I] $\rightarrow\infty$. However, if the dependence of λ on [I] is hyperbolic, DTI cannot be shortened below a given threshold because DTI will level off to a plateau even increasing [I] to infinity. Thus, for a modifier operating with mechanism M2, DTI$\rightarrow$ $5/(k_{-4} + k_4)$ for [I] $\rightarrow\infty$ and the success in the practical use of this modifier will depend on the values of k_{-4} and k_4.

These simple considerations emphasize that every effort put in determining mechanisms of action and kinetic parameters for slow-onset enzyme modification as accurately as possible is not an academic exercise. On the contrary, this knowledge is indispensable for predicting the physiological and pharmacological significance of the modifiers and can help in the chemical design of new drugs.

REFERENCES

[1] Frieden, C. (1970) Kinetic aspects of regulation of metabolic processes. The hysteretic enzyme concept. *J. Biol. Chem.* **245:**5788 – 5799.

[2] Cha, S. (1975) Tight-binding inhibitors – I. Kinetic behavior [Erratum: Biochem. Pharmacol. 25: 1561, 1976]. *Biochem. Pharmacol.* **24:**2177 – 2185. doi: 10.1016/0006-2952(75)90050-7.

[3] Cha, S. (1976) Tight-binding inhibitors – III. A new approach for the determination of competition between tight-binding inhibitors and substrates. Inhibition of adenosine deaminase by coformycin. *Biochem. Pharmacol.* **25:**2695 – 2702. doi: 10.1016/0006-2952(76)90259-8.

[4] Cha, S. (1980) Tight-binding inhibitors – VII. Extended interpretation of the rate equation. Experimental designs and statistical methods. *Biochem. Pharmacol.* **29:**1779 – 1789. doi: 10.1016/0006-2952(80)90140-9.

[5] Morrison, J.F. (1982) The slow-binding and slow, tight-binding inhibition of enzyme-catalysed reactions. *Trends Biochem. Sci.* **7:**102 – 105. doi: 10.1016/0968-0004(82)90157-8,

[6] Morrison, J.F., and Stone, S.R. (1985) Approaches to the study and analysis of the inhibition of enzymes by slow- and tight-binding inhibitors. *Comments Mol. Cell. Biophys.* **2:**347 – 368.

[7] Morrison, J.F., and Walsh, C.T. (1988) The behavior and significance of slow-binding inhibitors. *Adv. Enzymol. Relat. Areas Mol. Biol.* **61:**201 – 301.

[8]   Hijazi, N.H., and Laidler, K.J. (1973) Transient-phase and steady-state kinetics for enzyme activation. *Can. J. Biochem.* **51**:806 – 814.
      doi: 10.1139/o73-100.

[9]   Neet, K.E., Ohning, G.V., and Woodruff, N.R. (1984) Hysteretic enzymes, slow inhibition, slow activation, and slow membrane binding. In *Dynamics of biochemical systems*, J. Ricard, and A. Cornish-Bowden, eds. (New York, Plenum Press), pp. 3 – 28.

[10] Johnson, K.A., Simpson, Z. B., and Blom, T. (2009) FitSpace Explorer: An algorithm to evaluate multidimensional parameter space in fitting kinetic data. *Anal. Biochem.* **387**:30 – 41.
      doi: 10.1016/j.ab.2008.12.025.

[11] Johnson, K.A., Simpson, Z. B., and Blom, T. (2009) Global Kinetic Explorer: A new computer program for dynamic simulation and fitting of kinetic data. *Anal. Biochem.* **387**:20 – 29.
      doi: 10.1016/j.ab.2008.12.024.

[12] Walmsley, A.R., and Bagshaw, C.R. (1989) Logarithmic timebase for stopped-flow data acquisition and analysis. *Anal. Biochem.* **176**:313 – 318.
      doi: 10.1016/0003-2697(89)90315-1.

[13] Meh, P., Pavšič, M., Turk, V., Baici, A., and Lenarčič, B. (2005) Dual concentration-dependent activity of thyroglobulin type-1 domain of testican: specific inhibitor and substrate of cathepsin L. *Biol. Chem.* **386**:75 – 83.
      doi: 10.1515/BC.2005.010.

[14] Dunwiddie, C., Thornberry, N.A., Bull, H.G., Sardana, M., Friedman, P.A., Jacobs, J.W., and Simpson, E. (1989) Antistasin, a leech-derived inhibitor of factor Xa. Kinetic analysis of enzyme inhibition and identification of the reactive site. *J. Biol. Chem.* **264**:16694 – 16699.

[15] Duggleby, R.G., Attwood, P.V., Wallace, J.C., and Keech, D.B. (1982) Avidin is a slow-binding inhibitor of pyruvate carboxylase. *Biochemistry* **21**:3364 – 3370.
      doi: 10.1021/bi00257a018.

[16] Baici, A., and Gyger-Marazzi, M. (1982) The slow, tight-binding inhibition of cathepsin B by leupeptin. A hysteretic effect. *Eur. J. Biochem.* **129**:33 – 41.
      doi: 10.1111/j.1432-1033.1982.tb07017.x.

[17] Schultz, R.M., Varma-Nelson, P., Ortiz, R., Kozlowski, K.A., Orawski, A.T., Pagast, P., and Frankfater, A. (1989) Active and inactive forms of the transition-state analog protease inhibitor leupeptin: explanation of the observed slow binding of leupeptin to cathepsin B and papain. *J. Biol. Chem.* **264**:1497 – 1507.

[18]   Sivaraman, S., Zwahlen, J., Bell, A.F., Hedstrom, L., and Tonge, P.J. (2003) Structure-activity studies of the inhibition of FabI, the enoyl reductase from *Escherichia coli*, by triclosan: kinetic analysis of mutant FabIs. *Biochemistry* **42**:4406 – 4413. doi: 10.1021/bi0300229.

[19]   Novinec, M., Kovačič, L., Lenarčič, B., and Baici, A. (2010) Conformational flexibility and allosteric regulation of cathepsin K. *Biochem. J.* **429**:379 – 389. doi: 10.1042/BJ20100337.

[20]   Baici, A., Schenker, P., Wächter, M., and Rüedi, P. (2009) 3-Fluoro-2,4-dioxa-3-phosphadecalins as inhibitors of acetylcholinesterase. A reappraisal of kinetic mechanisms and diagnostic methods. *Chem. Biodivers.* **6**:261 – 282. doi: 10.1002/cbdv.200800334.

[21]   Wächter, M., and Rüedi, P. (2009) Synthesis and characterization of the enantiomerically pure cis- and trans-2,4-dioxa-3-fluoro-3-phosphadecalins as inhibitors of acetylcholinesterase. *Chem. Biodivers.* **6**:283 – 294. doi: 10.1002/cbdv.200800335.

[22]   Topham, C.M. (1990) A generalized theoretical treatment of the kinetics of an enzyme-catalysed reaction in the presence of an unstable irreversible modifier. *J. Theor. Biol.* **145**:547 – 572. doi: 10.1016/S0022-5193(05)80488-6.

[23]   Baici, A. (1988) Criteria for the choice of inhibitors of extracellular matrix-degrading endopeptidases. In *The control of tissue damage*, A.M. Glauert, ed. (Amsterdam, Elsevier), pp. 243 – 258.

[24]   Baici, A. (1998) Inhibition of extracellular matrix-degrading endopeptidases: Problems, comments, and hypotheses. *Biol. Chem.* **379**:1007 – 1018.

Beilstein-Institut

Experimental Standard Conditions of Enzyme Characterization
September 12th – 16th, 2011, Rüdesheim/Rhein, Germany

75

Ontology-based search in SABIO-RK

Ulrike Wittig[*], Enkhjargal Algaa, Andreas Weidemann, Renate Kania, Maja Rey, Martin Golebiewski, Lei Shi, Lenneke Jong and Wolfgang Müller

Scientific Databases and Visualization Group,
Heidelberg Institute for Theoretical Studies (HITS),
Schloss-Wolfsbrunnenweg 35, 69118 Heidelberg, Germany

E-Mail: *Ulrike.Wittig@h-its.org

Received: 31st January 2012/Published: 15th February 2013

Abstract

The SABIO-RK database (http://sabio.h-its.org/) is established as a resource for biochemical reactions and their kinetic data. Data are manually extracted from scientific literature and stored in a structured and standardised format. Additionally SABIO-RK allows direct submission of data from lab experiments in an automated workflow, e.g. within project collaborations for storage and exchange of unpublished experimental results and later publishing the data. To access the kinetic data in the database, web interfaces and web services are available offering complex searches using various criteria. For specific search criteria different classification levels of organisms, tissues, and reactants, can be selected based on biological ontologies. Biological ontologies are developed for a hierarchical classification of biological objects and for modelling a domain using shared vocabularies. Ontological relations are implemented in SABIO-RK to extend the search functionalities.

INTRODUCTION

SABIO-RK [1] is a manually curated database containing kinetic data of biochemical reactions and their related information. The data are manually extracted from literature [2] or are automatically submitted from lab experiments [3]. SABIO-RK combines available kinetic parameters along with their corresponding rate equations, as well as kinetic law and parameter types and experimental and environmental conditions (pH, temperature, buffer) under which the kinetic data were determined. It stores detailed information about the biochemical reactions and pathways including their reaction participants (substrates, products), modifiers (inhibitors, activators, cofactors), cellular location, enzyme information (e.g. UniProt [4] accession number, EC number [5], isozymes, protein complex composition, wild type/mutant information, molecular weight), kinetic parameters, corresponding rate equation, and biological source (organism, tissue, cell location). As data within SABIO-RK are strongly interlinked, they are mainly extracted from literature. [6] Each entry within SABIO-RK is referenced to the original publication and data directly submitted from lab experiments are referenced to the original source of the raw data (e.g. the MeMo-RK database storing the experimental raw data of our collaboration partners in Manchester). Biology and biochemistry experts are doing the curation and annotation to controlled vocabularies, ontologies and external data sources.

ONTOLOGIES AND CONTROLLED VOCABULARIES

A defined and shared vocabulary is important to avoid misinterpretations and helps to exchange data between resources correctly. Controlled vocabularies provide predefined terms for a specific domain, which have been selected by domain experts. In many cases these vocabularies are structured like taxonomies or hierarchical classifications and typically include so-called is_a and part_of relationships. To illustrate these relationships an example: Vehicles are cars, bicycles, trains and so on. Therefore the entities *car*, *bicycle* and *train* are denoted by *is_a relationships* to *vehicle*, even though all three have different characteristics. A car consists of several parts, including wheels or an engine, thus the entities *wheel* and *engine* have *part_of* relationships to *car*.

Ontologies, including controlled vocabularies and all relationships between objects, represent the knowledge of a well-defined domain and are used to describe properties of the objects.

Biological ontologies and controlled vocabularies are also being developed to force the usage of a shared language for naming of biological objects. By agreeing on a particular ontological representation, for example the development of biological databases, a common vocabulary can be used for the analysis of biological data and their comparison between different resources. In biology, chemistry or medicine, there are many synonymous terms, abbreviations and acronyms that can refer to the same object. For example *alpha-ketoglutarate* and *2-oxoglutarate* are synonymous names of the same chemical compound. For the identification of synonyms of chemical compounds our group developed ChemHits (http://sabio.h-its.org/chemHits/) as an analysis tool for chemical compound names.

With the increasing amount of data generated from high-throughput experiments, there is an increasing need for structuring the data. Ontologies assure the unambiguous identification of objects you are looking for and thus help to analyse the data correctly. Additionally ontologies also assist to build and maintain ontologies themselves. [7 – 9]

In biology the systematic definition of terms and hierarchical relations is very old. For the first time it was used to create a taxonomy for organisms. Organism names could change over time because new characteristics are described and the classification of the organism changes within the taxonomy: e. g. *Streptococcus faecalis* was used in the past and now it is changed to the genus *Enterococcus faecalis*. Because both organism names are used in publications a classification scheme and controlled vocabulary is needed including all synonyms. Another system for nomenclature and classification was established in the 1950's. At that time the number of newly discovered enzymes increased very rapidly. Now there was the need to unravel the different names for the same enzyme and to establish a system where enzymes with similar functions could be grouped in classes, leading to the EC classification system [5].

Most of the available biological ontologies and controlled vocabulary only contain hierarchical is_a relationships between objects. Based on these parent-child relations classifications and hierarchical structures of objects can be derived. Examples of such classifications are the enzyme classification system offered by IUBMB [5] or the organism taxonomy provided by NCBI [10]. There are also ontologies which have not only is_a, but also part_of relationships between objects. The BRENDA Tissue Ontology (BTO) [11] for example, contains information about tissues and cell types. The relationships between tissues and cell types are represented as part_of relationships, e. g. hepatocyte is part_of liver.

Beside these, the ChEBI ontology for small chemical compounds [12] is one of the few ontologies containing additional relationships, apart from is_a and part_of relations. In addition, the ChEBI ontology contains chemistry-specific relationships which can be used to convey additional information about the chemical compound. It includes acid-base relations, relations between different stereoisomeric forms of chemical compounds, or relations to define the functional role of a chemical compound.

To improve the functionality of the SABIO-RK database by offering extended queries hierarchies (is_a and part_of relations) based on the ontological relations from several ontologies are implemented in the SABIO-RK search options.

During the process of data insertion and curation controlled vocabularies are available for students extracting the data from literature and for database curators working on the data annotation and the quality control. For consistency and to avoid duplicate entries, for example lists of compounds, reactions, organisms, tissues, cellular locations, kinetic law types, parameter types, and units already existing in the SABIO-RK database are provided as selection lists for students and curators. Most of these lists are first generated by extracting terms from external sources and extended by terms extracted from literature. Enzyme names and EC numbers are extracted from IUBMB, organism names from NCBI taxonomy, tissues and cellular locations from BRENDA [13], types of kinetic laws and parameters from Systems Biology Ontology (SBO) [14], and units from the International System of Units [15]. Synonymic terms are referred to the same recommended term to enable the search for alternative names. New terms extracted from literature are added to the term lists and in many cases they were also submitted as new terms to the corresponding ontology. Therefore the SABIO-RK curators also help in the further development of external controlled vocabularies and ontologies (e. g. SBO or BTO).

Table 1. Overview of annotations and links in SABIO-RK to external databases, ontologies, and controlled vocabularies [4 – 5, 10 – 14, 18 – 23] and options how to search for them in SABIO-RK.

	Annotation to		Link to	Search by	
	Controlled vocabulary/ Ontology	Database		Name	ID
Reaction			KEGG	KEGG		X
Compound	ChEBI	ChEBI	ChEBI	X	X
		KEGG	KEGG	X	X
		PubChem	PubChem	X	
Enzyme	IUBMB	ExplorEnz	ExplorEnz		
		BRENDA	BRENDA		
		ExPASy	ExPASy		
		KEGG	KEGG		
		Reactome	Reactome		
		IntEnz	IntEnz		
			IUBMB	X	X
Protein		UniProt	UniProt		X
Organism	NCBI Taxonomy		NCBI Taxonomy	X	
Tissue	BTO		BTO	X	
Cell location			Gene Ontology	X	
Kinetic parameter type	SBO			X	
Kinetic law type	SBO			X	
Role in reaction	SBO			X	
Publication		PubMed	PubMed		X

Based on the known terms from selection lists data in SABIO-RK can be unambiguously identified and annotated to external data resources and ontologies. Biological ontologies and controlled vocabularies used for annotations in SABIO-RK are for example ChEBI, SBO, BTO, and NCBI taxonomy. In addition, comprehensive annotations to external databases enable the user to obtain further details, for example about reactions, compounds, enzymes, proteins, tissues, or organisms. A detailed listing of available annotations and links to databases and ontologies in SABIO-RK is represented in Table 1.

Annotations to external databases and ontologies together with the data from SABIO-RK can be also exported in SBML (Systems Biology Markup Language) [16], compliant with the MIRIAM standard [17], e.g. for setting up biochemical reaction network models. SABIO-RK reaction and kinetic law identifiers both are themselves listed as MIRIAM data types and are also exported with the data from allowing tracking of the data back to SABIO-RK as data source.

ONTOLOGY-BASED DATA SEARCH IN SABIO-RK

Data in SABIO-RK can be accessed via web-based user interfaces (classical and new interface) or via web-services (our previous SOAP slowly phasing out and being replaced by a more modern RESTful interface). Queries are available using names of biochemical objects or internal SABIO-RK or external database identifiers. Searching reactions using identifiers is offered for example by reaction identifiers from KEGG or SABIO-RK, by compound identifiers from KEGG [18], ChEBI, PubChem [10], or SABIO-RK, by enzyme EC numbers, UniProt accession numbers, or PubMed identifiers. Most of the search criteria can be defined by name searches, e.g. compounds, enzymes, organisms, tissues, cell locations, kinetic parameters, or kinetic law definitions (Table 1).

Some of the search criteria (e.g. organism, tissue, and compound) are based on biological ontologies. Based on different classification levels SABIO-RK can be searched not only for specific terms but also for a group of terms which have similar characteristics defined by subparts in the hierarchical tree of an ontology. These hierarchical relationships between objects extracted from is_a relations are implemented in the SABIO-RK search options. Therefore the search for organisms can be either defined by using a specific term, e.g. *Rattus norvegicus* or can be extended by defining a group of organisms based on the NCBI taxonomy, e.g. the search for all rodents including for example mouse or hamster by using the search term *Rodentia (NCBI)*. The selection lists contain single terms and terms with *NCBI* in parenthesis representing terms from the NCBI taxonomy. The search for NCBI taxonomy terms always includes all "children" and "grandchildren" of this term extracted from the NCBI taxonomy tree.

The tissue search in SABIO-RK includes the possibility to use BRENDA Tissue Ontology (BTO) terms. Therefore the search for tissues can be also defined by simple terms or terms extracted from the BTO. For example the search for *kidney* gives fewer search results compared to *kidney (BTO)* because the last one not only searches for the tissue *kidney* but also for several kidney cell lines or subparts of the tissue which are represented as is_a or part_of relationships in the ontology. In Figure 1 a screenshot of the SABIO-RK web interface represents the ontology-based search results for organism *Rodentia (NCBI)* and tissue *kidney (BTO)*.

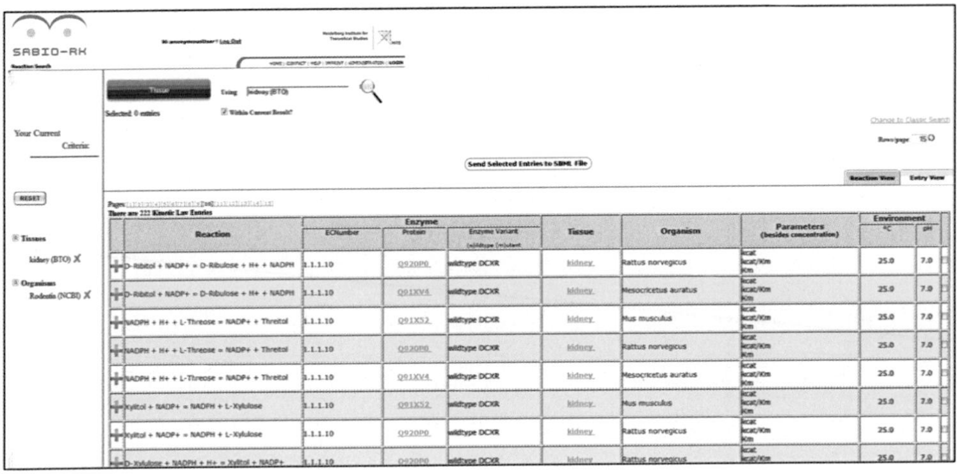

Figure 1. Screenshot of the SABIO-RK web interface for the search for *Rodentia (NCBI)* and *kidney (BTO)*.

At the moment only the classical SABIO-RK web-based interface offers for chemical compounds the search using ChEBI ontology terms. The is_a relationships extracted from the ChEBI sub-ontology "Molecular Structure" are implemented in the database search options for reaction participants to include the search for compound classes based on the hierarchical compound classification. For example queries can be defined to find all reactions containing an amino acid as reaction participant. Therefore for the query the reactant *amino acid* (CHEBI) has to be selected and the result would contain all reactions with any amino acid as substrate or product.

Based on the implementation of the hierarchical structures extracted from ontologies there is more functionality available so that the database user is able to decide which level of information is needed for the search. Queries including ontology-based terms result in more and comprehensive data. If a search result is dissatisfying an amplification of the search domain would help by using more general terms in the classification scheme. It also helps especially for tissue searches to get all related entries for one tissue including its cell lines because in the literature tissues and cell lines are described equivalent for similar experiments. For example in some publications the results are specified for liver but the experiments are done on hepatocytes which are specific cells of the liver. Ontology-based searches for tissues using BTO includes both tissues and corresponding cell lines in one query to offer a combination of related terms with the regard to the contents.

On the other hand ontology-based searches offer the possibility to easier compare kinetic data within for example groups of organisms with same characteristics. Therefore SABIO-RK offers with these new organism search options the comparison of all kinetic data of a group of organisms like for example plants, mammals, or vertebrates.

CONCLUSION

SABIO-RK is a curated database containing biochemical reactions and their kinetics. Extracted from some selected biological ontologies and controlled vocabularies the hierarchical relations are implemented in the SABIO-RK search options for advanced functionality of the database. Annotations in SABIO-RK to controlled vocabularies, ontologies, and external databases allow comprehensive searches in the database, linking to external sources and the comparison of data. The search for organisms, tissues, and compounds can be extended by the search using ontological terms from NCBI taxonomy, BRENDA Tissue Ontology, and ChEBI Ontology, respectively. Future work will include the extension of the ontology-based searches for other data in SABIO-RK, especially for cell locations.

ACKNOWLEDGEMENT

The SABIO-RK project is supported by the Klaus Tschira Foundation (http://www.klaus-tschira-stiftung.de/), the German Federal Ministry of Education and Research (http://www.bmbf.de/) through Virtual Liver and SysMO-LAB (Systems Biology of Microorganisms), and the DFG LIS (http://www.dfg.de/), under the short title "Integrated Immunoblot Environment".

REFERENCES

[1]   Wittig, U., Kania, R., Golebiewski, M., Rey, M., Shi, L., Jong, L., Algaa, E., Weidemann, A., Sauer-Danzwith, H., Mir, S., Krebs, O., Bittkowski, M., Wetsch, E., Rojas, I., Müller, W. (2012) SABIO-RK – database for biochemical reaction kinetics. *Nucleic Acids Res.* **40:**D 790 – 6. doi: 10.1093/nar/gkr1046.

[2] Wittig, U., Golebiewski, M., Kania, R., Krebs, O., Mir, S., Weidemann, A., Anstein, S., Saric, J., Rojas, I. (2006) SABIO-RK: integration and curation of reaction kinetics data. *Lecture Notes in Computer Science* **4075**:94 – 103.
doi: 10.1007/11799511_9.

[3] Swainston, N., Golebiewski, M., Messiha, H.L., Malys, N., Kania, R., Kengne, S., Krebs, O., Mir, S., Sauer-Danzwith, H., Smallbone, K., Weidemann, A., Wittig, U., Kell, D.B., Mendes, P., Müller, W., Paton, N.W., Rojas, I. (2010) Enzyme kinetics informatics: from instrument to browser. *FEBS Journal* **277**:3769 – 79.
doi: 10.1111/j.1742-4658.2010.07778.x.

[4] The UniProt Consortium (2011) Ongoing and future developments at the Universal Protein Resource. *Nucleic Acids Res.* **39**:D214 – 9.
doi: 10.1093/nar/gkq1020.

[5] IUBMB Enzyme Classification: http://www.chem.qmul.ac.uk/iubmb/enzyme/

[6] Wittig, U., Kania, R., Rojas, I., Müller, W. (2010) Herausforderungen bei der Extraktion von biochemischen Daten aus der Literatur. In *Lecture Notes in Informatics (LNI) – Proceedings, Series of the Gesellschaft für Informatik (GI).*

[7] Rojas, I., Ratsch, E., Saric, J., Wittig, U. (2004) Notes on the use of ontologies in the biochemical domain. *In Silico Biol.* **4**(1):89 – 96.

[8] Bodenreider, O., Stevens, R. (2006) Bio-ontologies: current trends and future directions. *Brief Bioinform.* **7**(3):256 – 74.
doi: 10.1093/bib/bbl027.

[9] Rubin, D.L., Shah, N.H., Noy, N.F. (2008) Biomedical ontologies: a functional perspective. *Brief Bioinform.* **9**(1):75 – 90.
doi: 10.1093/bib/bbm059.

[10] Sayers, E.W., Barrett, T., Benson, D.A., Bolton, E., Bryant, S.H., Canese, K., Chetvernin, V., Church, D.M., DiCuccio, M., Federhen, S., Feolo, M., Fingerman, I.M., Geer, L.Y., Helmberg, W., Kapustin, Y., Landsman, D., Lipman, D.J., Lu, Z., Madden, T.L., Madej, T., Maglott, D.R., Marchler-Bauer, A., Miller, V., Mizrachi, I., Ostell, J., Panchenko, A., Phan, L., Pruitt, K.D., Schuler, G.D., Sequeira, E., Sherry, S.T., Shumway, M., Sirotkin, K., Slotta, D., Souvorov, A., Starchenko, G., Tatusova, T.A., Wagner, L., Wang, Y., Wilbur, W.J., Yaschenko, E., Ye, J. (2011) Database resources of the National Center for Biotechnology Information. *Nucleic Acids Res.* **39**:D38 – 51.
doi: 10.1093/nar/gkq1172.

[11]   Gremse, M., Chang, A., Schomburg, I., Grote, A., Scheer, M., Ebeling, C., Schomburg, D. (2011) The BRENDA Tissue Ontology (BTO): the first all-integrating ontology of all organisms for enzyme sources. *Nucleic Acids Res.* **39**:D 507 – 13. doi: 10.1093/nar/gkq968.

[12]   de Matos, P., Alcántara, R., Dekker, A., Ennis, M., Hastings, J., Haug, K., Spiteri, I., Turner, S., Steinbeck, C. (2010) Chemical Entities of Biological Interest: an update. *Nucleic Acids Res.* **38**:D 249 – 54. doi: 10.1093/nar/gkp886.

[13]   Scheer, M., Grote, A., Chang, A., Schomburg, I., Munaretto, C., Rother, M., Söhngen, C., Stelzer, M., Thiele, J., Schomburg, D. (2011) BRENDA, the enzyme information system in 2011. *Nucleic Acids Res.* **39**:D 670 – 6. doi: 10.1093/nar/gkq1089.

[14]   Courtot, M., Juty, N., Knüpfer, C., Waltemath, D., Zhukova, A., Dräger, A., Dumontier, M., Finney, A., Golebiewski, M., Hastings, J., Hoops, S., Keating, S., Kell, D.B., Kerrien, S., Lawson, J., Lister, A., Lu, J., Machne, R., Mendes, P., Pocock, M., Rodriguez, N., Villeger, A., Wilkinson, D.J., Wimalaratne, S., Laibe, C., Hucka, M., Le Novère, N. (2011) Controlled vocabularies and semantics in systems biology. *Mol. Syst. Biol.* **7**:543. doi: 10.1038/msb.2011.77.

[15]   International System of Units (SI): http://www.bipm.fr/en/si/

[16]   Hucka, M., Finney, A., Sauro, H.M., Bolouri, H., Doyle, J.C., Kitano, H., Arkin, A.P., Bornstein, B.J., Bray, D., Cornish-Bowden, A. *et al.* (2003) The systems biology markup language (SBML): a medium for representation and exchange of biochemical network models. *Bioinformatics* **19**:524 – 31. doi: 10.1093/bioinformatics/btg015.

[17]   Le Novère, N., Finney, A., Hucka, M., Bhalla, U.S., Campagne, F., Collado-Vides, J., Crampin, E.J., Halstead, M., Klipp, E., Mendes, P., Nielsen, P., Sauro, H., Shapiro, B., Snoep, J.L., Spence, H.D., Wanner, B.L. (2005) Minimum Information Required In the Annotation of Models (MIRIAM). *Nat. Biotechnol.* **23**:1509 – 15. doi: 10.1038/nbt1156.

[18]   Kanehisa, M., Goto, S., Furumichi, M., Tanabe, M., Hirakawa, M. (2010) KEGG for representation and analysis of molecular networks involving diseases and drugs. *Nucleic Acids Res.* **38**:D 355 – 60. doi: 10.1093/nar/gkp896.

[19]   The Gene Ontology Consortium (2000) Gene ontology: tool for the unification of biology. *Nat. Genet.* **25**(1):25 – 9. doi: 10.1038/75556.

[20]   McDonald, A.G., Boyce, S., Moss, G.P., Dixon, H.B., Tipton, K.F. (2007) ExplorEnz: a MySQL database of the IUBMB enzyme nomenclature. *BMC Biochem.* **8**:14.
doi: 10.1186/1471-2091-8-14.

[21]   Bairoch, A. (2000) The ENZYME database in 2000. *Nucleic Acids Res.* **28**:304 – 5.
doi: 10.1093/nar/28.1.304.

[22]   Matthews, L., Gopinath, G., Gillespie, M., Caudy, M., Croft, D., de Bono, B., Garapati, P., Hemish, J., Hermjakob, H., Jassal, B., Kanapin, A., Lewis, S., Mahajan, S., May, B., Schmidt, E., Vastrik, I., Wu, G., Birney, E., Stein, L., D'Eustachio, P. (2009) Reactome knowledgebase of human biological pathways and processes. *Nucleic Acids Res.* **37**:D 619 – 22.
doi: 10.1093/nar/gkn863.

[23]   Fleischmann, A., Darsow, M., Degtyarenko, K., Fleischmann, W., Boyce, S., Axelsen, K.B., Bairoch, A., Schomburg, D., Tipton, K.F., Apweiler, R. (2004) IntEnz, the integrated relational enzyme database. *Nucleic Acids Res.* **32**:D 434 – 7.
doi: 10.1093/nar/gkh119.

Beilstein-Institut

Experimental Standard Conditions of Enzyme Characterization
September 12th – 16th, 2011, Rüdesheim/Rhein, Germany

85

Protein Engineering Elucidates the Relationship between Structure, Function and Stability of a Metabolic Enzyme

Reinhard Sterner, Thomas Schwab and Sandra Schlee

University of Regensburg, Institute of Biophysics and Physical Biochemistry, Universitätsstrasse 31, D-93053 Regensburg, Germany.

E-Mail: *Reinhard.Sterner@biologie.uni-regensburg.de

Received: 14th March 2012/Published: 15th February 2013

Abstract

The relationship between oligomerisation state, stability, and catalytic activity of the anthranilate phosphoribosyl transferase from *Sulfolobus solfataricus* (sAnPRT) was analysed by three interrelated protein engineering approaches. The extremely thermostable homodimeric sAnPRT enzyme was converted into a monomer by rational design, and its low catalytic activity at 37 °C was elevated by a combination of random mutagenesis and metabolic selection in the mesophilic host *Escherichia coli*. The two amino acid exchanges leading to monomerization and the two substitutions resulting in activation of sAnPRT were then combined, which resulted in an "activated monomer" that was significantly less stable and more active than wild-type sAnPRT. Using a combination of random mutagenesis and selection in the thermophilic host *Thermus thermophilus*, the activated monomer was stabilized, and the consequences of stabilization for catalytic activity and association state were analysed.

INTRODUCTION

It is important to understand how the tertiary and quaternary structure of an enzyme determines its thermal stability, catalytic activity, and conformational flexibility. Insights into the interrelationship between these properties were gained by the analysis of homologous enzymes from mesophiles and hyperthermophiles that are characterized by an optimum growth temperature close to the boiling point of water [1]. For example, crystal structure analysis suggests that the significantly higher stability of enzymes from hyperthermophiles compared to their counterparts from mesophiles is due to only minor modifications at the level of tertiary structure: a slightly elevated number of hydrogen bonds or salt bridges, a higher compactness, an increased polar to non-polar surface area, and raises in α-helix content and α-helix stability. At the level of quaternary structure, more pronounced differences have been observed occasionally: some enzymes from hyperthermophiles form higher-order oligomers than the homologous proteins from mesophiles [2, 3].

In natural enzymes, conformational integrity at high temperatures generally comes along with low catalytic activity at room temperature. This negative stability – activity correlation is understandable from a *biological* point of view: enzymes from hyperthermophiles do not need to catalyse reactions at 25 °C, and enzymes from mesophiles do not have to resist 100 °C in their natural habitats. From a *physicochemical* point of view, the pronounced differences in activity between mesophilic – thermophilic enzyme pairs when compared at identical temperatures must be due to subtle alterations in the protein chain, because the active site residues are conserved among homologous enzymes, independent of the growth optimum of the host organism [4]. Having in mind that enzymatic activity requires a certain degree of flexibility [5], it has been speculated that the low activity of thermostable enzymes at mesophilic temperatures is due to high conformational rigidity, which is relieved at elevated temperature. As a consequence, mesophilic – thermophilic enzyme pairs should be comparably flexible and active at the respective physiological conditions [6]. However, the general validity of this concept of 'corresponding states' [7] has remained under debate [8, 9].

Although interesting and attractive, the explanatory power of comparisons between homologous enzymes from mesophiles and hyperthermophiles is severely limited by the high degree of sequence diversity, which has accumulated in during the course of natural evolution. Even for a comparably high sequence conservation of 50% between a hypothetical pair of enzymes containing 200 residues, 100 amino acids are disparate. However, typically no more than about five residue substitutions will be responsible for differences in stability or catalysis, which can make their identification a laborious and time-consuming 'needle in the haystack' search.

An alternative approach for identifying minor structural modifications that cause substantial alterations of enzyme integrity and function is protein engineering, which can be performed either by rational design or directed laboratory evolution. Rational protein design approaches are generally based on high-resolution crystal structures that guide the introduction of amino acid exchanges at specific positions with preconceived effects on protein stability or activity [10]. The protein variants are produced by site-directed mutagenesis and purified, and the effects of the substitutions are analysed, typically by enzyme kinetics, unfolding studies, and structural analysis. In a directed evolution experiment amino acid exchanges are not introduced at predetermined positions but are incorporated at random sites, typically by error-prone PCR. From the resulting gene library, variants with elevated stability or activity are isolated by *in vivo* selection or high-throughput *in vitro* screening. Directed evolution often leads to surprising solutions with beneficial amino acid exchanges at unexpected positions, which makes this approach particularly instructive [11]. Importantly, variants generated by rational design or directed evolution typically carry only a few amino acid substitutions, a feature that considerably alleviates their analysis. Rational design and directed evolution experiments have shown that the increase of catalytic power comes at the expense of a reduced conformational stability, and *vice versa*. However, in a few cases, both enzymatic properties could be optimized simultaneously, supporting the notion that the trade-off between catalysis/flexibility and stability observed in natural proteins is not due to physicochemical constraints [12].

In the following, the results of protein engineering studies will be presented that were performed to elucidate the structure-function-stability relationship of anthranilate phosphoribosyl transferase from *Sulfolobus solfataricus* (sAnPRT), which is a hyperthermophilic archaeon with an optimum growth temperature of 80 °C. Phosphoribosyl transferases (PRTases) are involved in the metabolism of nucleotides and amino acids. They catalyse the Mg^{2+}-dependent displacement of pyrophosphate (PP$_i$) from 5'-phosphoribosyl-α1-pyrophosphate (PRPP) by a nitrogen-containing nucleophile, producing an α-substituted ribose-5-phosphate. On the basis of their tertiary structures, PRTases have been divided into three different classes. Members of class I are the orotate and uracil PRTases as well as the purine PRTases. Representatives of class II are the quinolinate and nicotinic acid PRTases [13]. The only known member of class III is anthranilate phosphoribosyltransferase (AnPRT), which is encoded by the *trp*D gene and catalyses the third step of tryptophan biosynthesis, the ribosylation of anthranilate to phosphoribosyl anthranilate (PRA) (Figure 1).

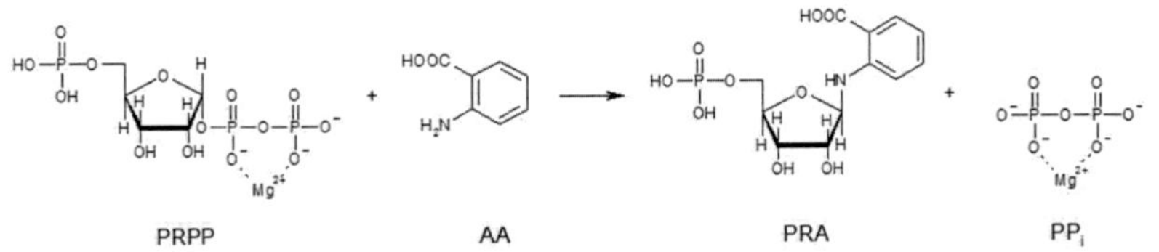

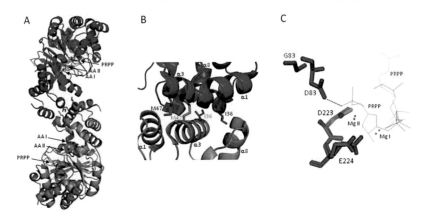

Figure 1. Reaction catalysed by AnPRT. PRPP, 5'-phophoribosyl-α1-pyrophosphate; AA, anthranilate; PRA, *N*-(5'-phosphoribosyl)anthranilate; PP$_i$, pyrophosphate.

RESULTS

Monomerisation of homodimeric sAnPRT by rational protein design

Figure 2A shows a ribbon diagram of the structure of sAnPRT, which was solved in complex with two molecules of anthranilate (AA) and one molecule of PRPP [14].

Figure 2. Crystal structures of sAnPRT. **(A)** Ribbon diagram of dimeric sAnPRT with bound substrates anthranilate (AA I and AA II) and PRPP. **(B)** Dimer interface of sAnPRT formed by helices α1, α3, and α8. Residues I36 and M47, which are located at the N- and C-termini of helix α3, are displayed as sticks. **(C)** Plot of PRPP and metal coordination by wild-type sAnPRT (blue) and sAnPRT-D83G+F149S (red). The structures have been superimposed on the main chain atoms of the conserved acidic motif (D223-E224). The hydrogen bond between the 5'-phosphate group of PRPP with the carboxyl side chain of aspartate 83 is indicated by a dashed line.

The enzyme forms a homodimer of identical subunits that consist of two domains: a large α/β domain, formed by a central β-sheet together with a C-terminal cluster of eight helices, and a small α-helical domain, comprising six helices. The substrate binding cavity for the coordination of two molecules of AA, one molecule of PRPP, and two Mg^{2+}-ions (Mg-I and Mg-II) is located at the domain interface, while dimer formation is mediated by the small α-helical domains which associate in a head-to-head fashion displaying an approximate two-fold symmetry. Wild-type sAnPRT is extremely thermostable with a half-life at 85 °C of 35 min [15] and catalytically proficient at 60 °C with a turnover number of $4.2 \, s^{-1}$ [14], but only marginally active at 37 °C with a turnover number of $0.33 \, s^{-1}$.

In order to elucidate the significance of homo-dimer formation for stability and activity of sAnPRT, the enzyme was monomerised by performing rational design on the basis of the crystal structure of the enzyme. Residues crucial for dimer formation were identified by inspecting the interface of the two protomers, which is mainly formed by residues from helices α1, α3 and α8 (Figure 2B). An *in silico* analysis suggested that the hydrophobic residues Ile36 and Met47, which are located at the N- and C-termini of helix α3, form the most numerous and intimate reciprocal inter-subunit interactions. Based on this finding, site-directed mutagenesis was used to replace Ile36 and Met47 by the acidic residues glutamate and aspartate, both individually and in combination. We reasoned that the introduction of the negatively charged side chains would weaken the stabilizing interactions of Ile36 and Met47 with residues of the other subunit. Moreover, the relatively low distances between the C_β-atoms of the symmetry-related Ile36-Ile36' (6.8 Å) and Met47-Met47' (4.4 Å) residue pairs indicated that the introduced negative charges further weaken inter-subunit interactions by electrostatic repulsion. Moreover, negatively charged residues located at the surface increase protein solubility, which could stabilize the monomer. Met47 was replaced by aspartate instead of glutamate, because its shorter side chain has a lower probability of forming a stabilizing hydrogen bond with Lys13' from the neighbouring protomer.

The double variant sAnPRT-I36E+M47D and the two single variants sAnPRT-I36E and sAnPRT-M47D were produced in *E. coli* and purified, and their association state was assessed by gel filtration chromatography and analytical ultracentrifugation. The results demonstrated that the double variant formed a homogeneous monomer, whereas the single variants were present in a dimer-monomer equilibrium. Steady-state enzyme kinetics per-formed at 37 °C showed that sAnPRT-I36E+M47D has, within experimental error, an iden-tical turnover number (k_{cat}) as the wild-type enzyme and unaltered Michaelis constants for AA (K_m^{AA}) and PRPP (K_m^{PRPP}). In contrast, differential scanning calorimetry and thermal inactivation revealed that the melting temperature (T_M) of sAnPRT-D83G+F149S was decreased by ~ 10 °C and that its inactivation at 80 °C ($t_{1/2(80 \, °C)}$) was accelerated ~ 10-fold. In order to discriminate between destabilization caused by the substitutions *per se* from destabilization caused by monomerisation, inactivation kinetics were performed in presence of various protein concentrations. At the lowest applied subunit concentration, sAnPRT-I36E and sAnPRT-M47D were present as monomers, and their $t_{1/2(80°C)}$ values of 4 min and 3 min were identical to the concentration-independent half-life of sAnPRT-I36E+M47D. At the highest tested concentrations, sAnPRT-I36E and sAnPRT-M47D became completely or partly dimeric, and their $t_{1/2(80°C)}$ values of about 40 min and 15 min approached the con-centration-independent half-life of wild-type sAnPRT.

These results prove that the introduced amino acid substitutions did not influence inactivation kinetics significantly, and that the higher stability of wild-type sAnPRT compared to sAnPRT-I36E+M47D is mainly due to dimerisation. We concluded that the monomeric double variant of sAnPRT resembles the experimentally inaccessible isolated wild-type subunits, and that sAnPRT is a dimer for reason of stability but not activity [16].

Activation of sAnPRT by library selection in Escherichia coli

The goal of this study was to increase the low catalytic activity of sAnPRT at 37 °C, and to assess the consequences of activation for thermal stability. Although the crystal structure with bound anthranilate, PRPP, and Mg^{2+} allowed for the identification of individual residues involved in substrate binding, it provided no hint as to which amino acid exchanges might lead to the activation of the enzyme. Given the absence of a rationale for pre-assigned substitutions, we turned to a combination of random mutagenesis and selection *in vivo*. For this purpose, the *strp*D gene encoding sAnPRT was amplified by error-prone PCR under conditions that led to the introduction of an average of $2-3$ amino acid exchanges of the protein. The amplification products were ligated into a plasmid that allows for gene expression in *E. coli*. The ligation mixture was used to transform an *E. coli* Δ*trp*EGD strain lacking the first three genes of the tryptophan biosynthetic pathway, which encode the anthranilate synthase complex (*trp*E and *trp*G code for the large and small subunit, respectively) and AnPRT (*trp*D). For growth on medium plates without tryptophan, Δ*trp*EGD requires externally added anthranilate and transformation with a *trp*D gene that codes for an AnPRT with significant catalytic activity at 37 °C. Whereas Δ*trp*EGD cells transformed with wild-type *strp*D needed 80 h to form colonies on minimal medium supplemented with anthranilate, a number of cells transformed with the plasmid mixture containing randomized *strp*D grew to a visible size within $16-48$ h. The faster growing colonies were expected to produce sAnPRT variants with a higher catalytic activity than the wild-type enzyme at 37 °C, which is more than 40 degrees below the physiological temperature of *S. solfataricus*. Sequencing of the *strp*D inserts isolated from the fastest growing colonies revealed mutations that resulted in the exchange of aspartate 83 by glycine (D 83G) or asparagine (D 83N), and the substitution of phenylalanine 149 by serine (F 149S).

The double variants AnPRT-D 83G+F 149S and the two single variants sAnPRT-D 83G and sAnPRT-F 149S were produced in *E. coli* and purified. Analytical gel filtration chromatography showed that all three variants retained the homodimeric association state of wild-type sAnPRT. Steady-state enzyme kinetics performed at 37 °C demonstrated that the pronounced inhibition of wild-type sAnPRT by high concentrations of Mg^{2+} was no longer present in sAnPRT-D 83G+F 149S and sAnPRT-D 83G. In order to explain this observation, sAnPRT-D 83G+F 149S was crystallized in presence of PRPP and Mn^{2+}, which can be identified more reliably in electron density maps than Mg^{2+}. The analysis of the X-ray structure revealed that PRPP bound to sAnPRT-D 83G+F 149S adopts an extended conformation that contrasts markedly with the "S" compact shape observed in complexes of wild-type sAnPRT

(Figure 2C). The "S" shape of PRPP in the wild-type enzyme is stabilized by a hydrogen bond between the 5'-phosphate group of the substrate with the carboxyl side chain of aspartate 83. Within sAnPRT-D83G+F149S, the hydrogen bond is lost as a result of the D83G exchange. For wild-type sAnPRT, the 5'-phosphate group of PRPP is involved in the binding of the second Mg^{2+}-ion (Mg-II). Although a second bivalent cation is also present in sAnPRT-D83G+F149S, the 5'-phosphate group of the extended PRPP is too far away to contribute to its binding. We therefore speculated that the affinity of Mg-II for the active site might be decreased in sAnPRT-D83G+F149S and thus the propensity for the transformation of the productive PRPP*Mg-I complex into a putatively inhibitory Mg-II*PRPP*Mg-I complex might be lower in the double variant than in wild-type sAnPRT. This assumption was confirmed by the thorough analysis of the PRPP- and Mg^{2+}-dependent enzymatic activities of wild-type sAnPRT and sAnPRT-D83G+F149S [17].

In addition to abolishing inhibition by Mg^{2+}, the two amino acid substitutions activate wild-type sAnPRT by a 40-fold enhancement of the turnover number at 37 °C in presence of the respective optimum concentrations of Mg^{2+} (wild-type sAnPRT: $k_{cat} = 0.33$ s^{-1}, 37 °C, 50 µM Mg^{2+}; sAnPRT-D83G+F149S: $k_{cat} = 13.3$ s^{-1}, 37 °C, 2 mM Mg^{2+}). Pre-steady state kinetic measurements were performed to determine the rate-limiting step of the AnPRT reaction, which – according to the differences in k_{cat} – is accelerated 40-fold for sAnPRT-D83G+F149S compared to wild-type sAnPRT. The minimal kinetic reaction scheme used to analyse the data is shown in Figure 3.

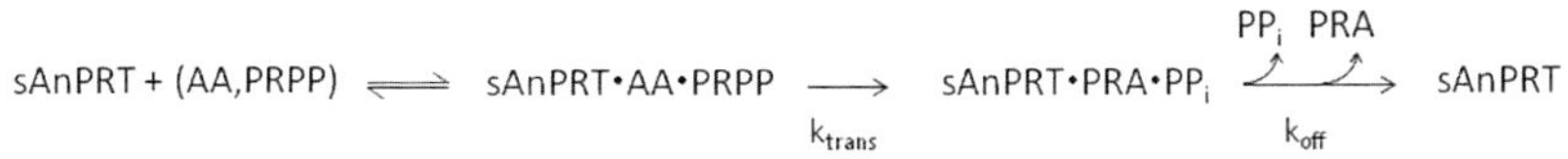

$$\text{sAnPRT} + (\text{AA,PRPP}) \rightleftharpoons \text{sAnPRT·AA·PRPP} \xrightarrow{k_{trans}} \text{sAnPRT·PRA·PP}_i \xrightarrow[k_{off}]{\text{PP}_i \quad \text{PRA}} \text{sAnPRT}$$

Figure 3. Minimal catalytic mechanism of the sAnPRT reaction with first-order rate constants describing the chemical transfer step (k_{trans}) and product release (k_{off}).

After the formation of the sAnPRT*AA*PRPP complex, a phosphoribosyl moiety is transferred from PRPP to AA (described by k_{trans}), followed by the release of the products PRA and PP$_i$ (described by k_{off}). The turnover number measured in the steady-state is determined by the chemical transfer and product release steps as follows: $k_{cat} = (k_{trans} \times k_{off})/(k_{trans} + k_{off})$. Therefore, the increase in k_{cat} observed for sAnPRT-D83G+F149S could be caused by an increase of either k_{trans} or k_{off}, depending on whether product release is comparably fast and chemical transfer is rate-limiting ($k_{trans} < k_{off}$), or vice versa ($k_{trans} > k_{off}$). To distinguish between the two possibilities, the turnover numbers of wild-type sAnPRT and sAnPRT-D83G+F149S determined by steady-state kinetics were compared with the value of k_{trans} as determined from transient kinetic experiments performed under single turnover conditions in a stopped-flow apparatus. For this purpose, a pre-incubated solution containing AA and an excess of sAnPRT was rapidly mixed with saturating concentrations of PRPP in presence of the respective optimal concentration of Mg^{2+}. Under conditions where the formation of sAnPRT*AA*PRPP is complete within the dead-time of the experiment (sAnPR-

T*AA*PRPP $\approx$ [A]$_{total}$; single-turnover conditions), the reaction is described by a two-step irreversible process (Figure 3). Since the spectroscopic change occurs upon product formation at the active site of the enzyme, the observed first-order rate constant k_{obs} corresponds to k_{trans}. An analysis of the dependence of k_{obs} on enzyme concentration yielded $k_{trans} = 29.6 \pm 0.9\,s^{-1}$ for sAnPRT-D83G+F149S, and $k_{trans} > 3.3 \pm 0.16\,s^{-1}$ for sAnPRT. From the relationship $k_{cat} = (k_{trans} \times k_{off})/(k_{trans} + k_{off})$, it follows that $k_{off} = 24.2 \pm 2.0\,s^{-1}$ for the double variant and $k_{off} \sim 0.33\,s^{-1}$ for the wild-type enzyme. For sAnPRT-D83G+F149S $k_{trans} \sim k_{off}$, demonstrating that phosphoribosyl transfer and product release affect catalytic turnover equally. For wild-type sAnPRT $k_{trans} > k_{off} = k_{cat}$, showing that product release determines the overall catalytic rate. In summary, the comparison of steady-state and single turnover data shows that the increased turnover number of sAnPRT-D83G+F149S is mainly based on the accelerated liberation of product. Since dissociation of PP$_i$ has been shown to be fast in other PRTases [18], the two amino acid exchanges appear to facilitate the release of PRA from the enzyme.

Differential scanning calorimetry and thermal inactivation revealed that the melting temperature of sAnPRT-D83G+F149S was decreased by ~10 °C and its inactivation at 80 °C was accelerated ~10-fold. The analysis of the two single variants showed that this destabilization is entirely caused by the F149S exchange. Remarkably, F149 is located within the hinge between the two domains of the sAnPRT protomer. One can speculate that a substitution at this position has an influence on the dynamics of *bona fide* domain opening and closure movements, which might be required for the release of PRA. Along these lines, it has been shown for glutathione transferase that an increase in local flexibility can accelerate the rate-limiting release of product in the millisecond time range (for sAnPRT-D83G+F149S: $k_{off} = 25\,s^{-1}$, $t_{1/2} = 28\,ms$), a time scale that is similar to that expected for the upper limit of large amplitude segmental motions [19]. For sAnPRT-wt, the k_{cat} value increases from $0.33\,s^{-1}$ at 37 °C to $4.2\,s^{-1}$ at 60 °C [14], suggesting that the velocity of product release, which limits the activity of the parental protein at 37 °C, is accelerated at 60 °C by a similar increase in flexibility as by the F149S replacement at 37 °C [17].

Stabilisation of the activated sAnPRT monomer by library selection in *Thermus thermophilus*

We were interested to assess the effect of combining the amino acid exchanges leading to monomerisation (I36E, M47D) and activation (D83G, F149S) of sAnPRT. For this purpose, the *strp*D-I36E+M47D+D83G+F149S gene was generated by site-directed mutagenesis and expressed in *E. coli*. The recombinant sAnPRT-I36E+M47D+D83G+F149S protein ("activated monomer") was purified and characterized. Steady-state enzyme kinetics and thermal unfolding using DSC and far-UV CD spectroscopy yielded a k_{cat} at 37 °C of $3.4\,s^{-1}$ and a T$_M$ of ~70 °C. Consequently, the turnover number is 10-fold higher and the melting temperature 20 °C lower compared to wild-type sAnPRT. The results show that the activating effect of the D 83G and F149S exchanges is somewhat less pronounced in the monomer than in the dimer, and that the destabilizing effects caused by monomerization and activation (both modifications lead to a decrease of T_M by ~10 °C) sum up in the activated monomer.

We attempted to "re-stabilize" sAnPRT-I36E+M47D+D 83G+F149S by a combination of random mutagenesis and metabolic selection in *Thermus thermophilus*. This extremely thermophilic bacterium is an ideal host for protein stabilization by library selection, because it grows between 55 °C and 80 °C and has a high natural competence for DNA uptake. Due to these properties, *T. thermophilus* cells transformed with a gene library can be incubated at elevated temperatures where survival of the host depends on the stabilization of the target protein. Taking advantage of the high transformation efficiency of *E. coli*, we used the mesophilic host for the generation of a large plasmid-encoded library of the *strpD*-I36E+M47D+D 83G+F149S gene. For this purpose, the gene was amplified by error-prone PCR, and the resulting repertoire was ligated into an *E. coli*-*T. thermophilus* shuttle plasmid. Transformation of highly competent *E. coli* cells yielded a library consisting of 9×10^6 individual plasmid-encoded mutants carrying a various number of nucleotide exchanges.

The shuttle plasmids were isolated and used to transform a *T. thermophilus* Δ*trpD* strain, which requires for growth in the absence of tryptophan a *trpD* gene encoding an AnPRT with high thermal stability. Control experiments showed that Δ*trpD* cells transformed with *strpD*-I36E+M47D+D 83G+F149S allowed for growth on selective medium up to only 70 °C, whereas transformation with wild-type *strpD* yielded colonies up to 79 °C. A number of Δ*trpD* cells transformed with the plasmid library were also able to grow at 79 °C, indicating that they contained variants with a higher thermal stability than sAnPRT-I36E+M47D+D 83G+F149S. Sequencing of the *strpD* inserts isolated from 20 colonies revealed that the selected variants contained three amino acid exchanges on average. For further analysis, we focused on the T77I, N109S, I169T, F193S and L 320 M substitutions, which either occurred as single exchange in a selected variant or were found independently in at least two different variants. The exchanges were introduced individually into the *strpD*-I36E+M47D+D 83G+F149S gene by site-directed mutagenesis, and the recombinant proteins were produced in *E. coli*, purified, and characterized. Thermal unfolding revealed that the T77I and F193S substitutions resulted in considerable T_M-increases of approximately 8 °C and 4 °C, whereas each of the N109S, I169T and L 320 M exchanges lead to a stabilization by only about 1 °C. The combination of the beneficial mutations showed that the T_M increases were additive to a first approximation. As a consequence, the variant containing all five identified stabilizing exchanges displayed a T_M of about 83.5 °C, which is 13 °C higher than the melting temperature of sAnPRT-I36E+M47D+D 83G+F149S (Figure 4).

Sterner, R. *et al.*

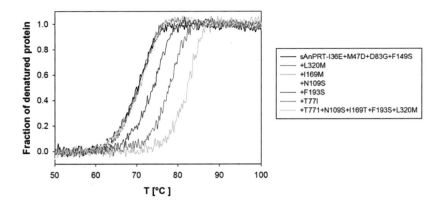

Figure 4. Stabilizing effect of amino acid exchanges identified by library selection in *T. thermophilus*. Thermal unfolding traces were recorded by loss of the far-UV CD signal at 220 nm.The staring construct sAnPRT-I36E+M47D+D83G+F149S has aT_M-value of 70.5 °C, which is increased to 83.5 °C by the combined substitutions T77I+N109S+I169T+F193S+L 320 M.

The T77I, N109S, I169T, F193S and L 320 M substitutions are distributed all over the structure of sAnPRT. Computational modelling suggests that replacing the polar side chain of threonine with the larger hydrophobic side chain of isoleucine (T77I) fills a cavity in the protein interior, explaining the strong increase of thermal stability caused by this exchange. In the absence of a high resolution crystal structure, the basis for the stabilizing effect of the other, less beneficial substitutions has remained unclear so far. In any case, analytical gel filtration showed that all variants were monomeric, demonstrating that stabilization was not due to "re-dimerization" of sAnPRT.

The catalytic activity of sAnPRT-I36E+M47D+D83G+F149S and its stabilized variants was measured by steady-state enzyme kinetics at 37 °C. As outlined in the previous section, the D83G exchange leads to the removal of the inhibition of wild-type sAnPRT by high concentrations of Mg^{2+}. Accordingly, sAnPRT-I36E+M47D+D83G+F149S and its variants are maximally active in the presence of millimolar concentrations of the bivalent cation. The two strongly stabilizing exchanges T77I and F193S modestly reduce the k_{cat} of sAnPRT-I36E+M47D+D83G+F149S by a factor of 2.5 and 1.3, respectively. Activity is stronger affected by the N109S exchange, which results in a 4.5-fold reduction of k_{cat}, probably because it is located close to the active site. The two stabilizing mutations I169T and L 320 M do not significantly alter the turnover number.

In order to obtain a comprehensive picture about the stability-activity relationship, the k_{cat}-values at 37 °C of all characterized monomeric sAnPRT variants were plotted as a function of the respective T_M (Figure 5).

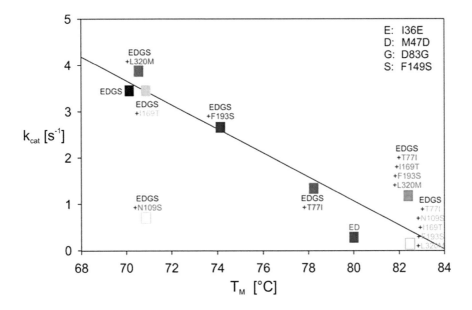

Figure 5. Negative correlation between turnover number and thermal stability of monomeric sAnPRT variants. The k_{cat} values at 37 °C were plotted as function of T_M. A negative linear correlation ($r^2 = 0.88$) is observed. Variants carrying the N109S substitution were excluded from the analysis, because this exchange is located close to the active site.

The observed negative correlation is in line with the view that protein flexibility is important for sAnPRT catalysis [20].

CONCLUSIONS

The three projects presented in this contribution underline the relevance of protein engineering for elucidating the relationship between the quaternary structure, the catalytic activity and the thermodynamic stability of metabolic enzymes. The non-covalent association of polypeptide chains can have various consequences for enzyme function, for example it can allow for the formation of complex sites or enable substrate channelling [21, 22]. However, the monomerization of sAnPRT [16] and a similar study performed with the homodimeric phosphoribosyl anthranilate isomerase from the hyperthermophile *Thermotoga maritima* [23] show that these enzymes are dimers for stability but not for activity reasons. Whereas the monomerization of sAnPRT was based on the analysis of the crystal structure of the protein, our limited understanding of enzyme-catalysed phosphoribosyl transfer did not allow for a rationale approach to increase the weak activity of sAnPRT at low temperatures. We therefore used directed evolution including random mutagenesis and selection in a mesophilic host to achieve this goal. This approach led to the identification of two *a priori* unpredictable activating amino acid exchanges whose effects on catalysis and stability were comprehensively analysed. An analogous directed evolution approach, which however used a thermophilic instead of a mesophilic host for selection, allowed for the stabilisation of the activated monomer. Remarkably, none of the identified exchanges led to the re-dimerisation of sAnPRT, confirming the finding that an increased association state is only one of numerous mechanisms to stabilize a protein [2]. The analysis of the activated and the stabilized variants suggest that conformational flexibility of sAnPRT required for efficient catalysis goes at the cost of thermal stability. These findings underline the notion that protein engineering approaches can be highly instructive provided that the isolated variants are subjected to a detailed characterisation by protein chemistry, enzyme kinetics, and crystal structure analysis.

REFERENCES

[1]   Stetter, K.O. (2006) History of discovery of the first hyperthermophiles. *Extremophiles* **10**:357 – 362.
       doi: 10.1007/s00792-006-0012-7.

[2]   Sterner, R. and Liebl, W. (2001) Thermophilic adaptation of proteins. *Crit. Rev. Biochem. Mol. Biol.* **36**:39 – 106.
       doi: 10.1080/20014091074174.

[3]   Vieille, C. and Zeikus, G.J. (2001) Hyperthermophilic enzymes: Sources, uses, and molecular mechanisms for thermostability. *Microbiol. Mol. Biol. Rev.* **65**:1 – 43.
       doi: 10.1128/MMBR.65.1.1-43.2001.

[4]   Jaenicke, R. and Böhm, G. (1998) The stability of proteins in extreme environments. *Curr. Opin. Struct. Biol.* **8**:738 – 748.
       doi: 10.1016/S0959-440X(98)80094-8.

[5]   Benkovic, S.J. and Hammes-Schiffer, S. (2003) A perspective on enzyme catalysis. *Science* **301**:1196 – 1202.
       doi: 10.1126/science.1085515.

[6]   Zavodszky, P., Kardos, J., Svingor, and Petsko, G.A. (1998) Adjustment of conformational flexibility is a key event in the thermal adaptation of proteins. *Proc. Natl. Acad. Sci. U.S.A.* **95**:7406 – 7411.
       doi: 10.1073/pnas.95.13.7406.

[7]   Jaenicke, R. (1991) Protein stability and molecular adaptation to extreme conditions. *Eur. J. Biochem.* **202**:715 – 728.
       doi: 10.1111/j.1432-1033.1991.tb16426.x.

[8]   Hernandez, G., Jenney, F.E., Jr., Adams, M.W., and LeMaster, D.M. (2000) Milli-
        second time scale conformational flexibility in a hyperthermophile protein at ambient
        temperature. *Proc. Natl. Acad. Sci. U.S.A.* **97**:3166 – 3170.
        doi: 10.1073/pnas.97.7.3166.

[9]   Jaenicke, R. (2000) Do ultrastable proteins from hyperthermophiles have high or low
        conformational rigidity? *Proc. Natl. Acad. Sci. U.S.A.* **97**:2962 – 2964.
        doi: 10.1073/pnas.97.7.2962.

[10]   Eijsink, V.G., Bjork, A., Gaseidnes, S., Sirevag, R., Synstad, B., van den Burg, B.,
        and Vriend, G. (2004) Rational engineering of enzyme stability. *J. Biotechnol.*
        113:105 – 120.
        doi: 10.1016/j.jbiotec.2004.03.026.

[11]   Bershtein, S. and Tawfik, D. S. (2008) Advances in laboratory evolution of enzymes.
        Curr. Opin. Chem. Biol. **12**:151 – 158.
        doi: 10.1016/j.cbpa.2008.01.027.

[12]   Sterner, R. and Brunner, E. (2008) Relationships among catalytic activity, structural
        flexibility, and conformational stability as deduced from the analysis of mesophilic-
        thermophilic enzyme pairs and protein engineering studies, In *Thermophiles – Biol-
        ogy and technology at high temperatures* (Robb, F., Antranikian, G., Grogan, D., and
        Driessen, A., Eds.), pp 25 – 28, CRC press, Taylor & Francis Group, LLC, Boca
        Raton.

[13]   Sinha, S.C. and Smith, J.L. (2001) The PRT protein family. *Curr. Opin. Struct. Biol.*
        11:733 – 739.
        doi: 10.1016/S0959-440X(01)00274-3.

[14]   Marino, M., Deuss, M., Svergun, D.I., Konarev, P.V., Sterner, R., and Mayans, O.
        (2006) Structural and mutational analysis of substrate complexation by anthranilate
        phosphoribosyltransferase from *Sulfolobus solfataricus. J. Biol. Chem.* **281**:21410 –
        21421.
        doi: 10.1074/jbc.M601403200.

[15]   Ivens, A., Mayans, O., Szadkowski, H., Wilmanns, M., and Kirschner, K. (2001)
        Purification, characterization and crystallization of thermostable anthranilate phos-
        phoribosyltransferase from *Sulfolobus solfataricus. Eur. J. Biochem.* **268**:2246 –
        2252.
        doi: 10.1046/j.1432-1327.2001.02102.x.

[16]   Schwab, T., Skegro, D., Mayans, O., and Sterner, R. (2008) A rationally designed monomeric variant of anthranilate phosphoribosyltransferase from *Sulfolobus solfataricus* is as active as the dimeric wild-type enzyme but less thermostable. *J. Mol. Biol.* **376**:506 – 516.
doi: 10.1016/j.jmb.2007.11.078.

[17]   Schlee, S., Deuss, M., Bruning, M., Ivens, A., Schwab, T., Hellmann, N., Mayans, O., and Sterner, R. (2009) Activation of anthranilate phosphoribosyltransferase from *Sulfolobus solfataricus* by removal of magnesium inhibition and acceleration of product release. *Biochemistry* **48**:5199 – 5209.
doi: 10.1021/bi802335s.

[18]   Wang, G.P., Lundegaard, C., Jensen, K.F., and Grubmeyer, C. (1999) Kinetic mechanism of OMP synthase: A slow physical step following group transfer limits catalytic rate. *Biochemistry* **38**:275 – 283.
doi: 10.1021/bi9820560.

[19]   Codreanu, S.G., Ladner, J.E., Xiao, G., Stourman, N.V., Hachey, D.L., Gilliland, G.L., and Armstrong, R.N. (2002) Local protein dynamics and catalysis: detection of segmental motion associated with rate-limiting product release by a glutathione transferase. *Biochemistry* **41**:15161 – 15172.
doi: 10.1021/bi026776p.

[20]   Schwab, T. and Sterner, R. (2011) Stabilization of a metabolic enzyme by library selection in *Thermus thermophilus*. *ChemBioChem* **12**:1581 – 1588.
doi: 10.1002/cbic.201000770.

[21]   Miles, E.W., Rhee, S. and Davies, D.R. (1999) The molecular basis of substrate channeling: *J. Biol. Chem.* **274**:12193 – 12196.
doi: 10.1074/jbc.274.18.12193.

[22]   Raushel, F.M., Thoden, J.B., and Holden, H.M. (2003) Enzymes with molecular tunnels: *AccChemRes* **36**:539 – 548.
doi: 10.1021/ar020047k.

[23]   Thoma, R., Hennig, M., Sterner, R., and Kirschner, K. (2000) Structure and function of mutationally generated monomers of dimeric phosphoribosylanthranilate isomerase from *Thermotoga maritima*. *Structure Fold. Des.* **8**:265 – 276.
doi: 10.1016/S0969-2126(00)00106-4.

Beilstein-Institut

Experimental Standard Conditions of Enzyme Characterization
September 12th – 16th, 2011, Rüdesheim/Rhein, Germany

99

Successes and Challenges in Functional Assignment in a Superfamily of Phosphatases

Karen N. Allen[1,*] and Debra Dunaway-Mariano[2,#]

[1]Department of Chemistry, Boston University, 590 Commonwealth Avenue, Boston, MA 02215 – 2521, U.S.A.

[2]Department of Chemistry and Chemical Biology, University of New Mexico, Albuquerque, NM, 87131, U.S.A.

E-Mail: *drkallen@bu.edu and #dd39@unm.edu

Received: 17th September 2012/Published: 15th February 2013

Abstract

The explosion of protein sequence information from genome sequencing efforts requires that current experimental strategies for function assignment must evolve into computationally-based function prediction. This necessitates the development of new strategies based, in part, on the identification of sequence markers, including residues that support structure and specificity as well as a more informed definition of orthologues. We have undertaken the function assignment of unknown members of the haloalkanoate dehalogenase superfamily using an integrated bioinformatics/structure/mechanism approach. Notably, a number of members show "substrate blurring", with activity toward a number of substrates and significant substrate overlap between paralogues. Other family members have been honed to a specific substrate with high catalytic efficiency and proficiency. Our findings highlight the use of the cap domain structure and enzyme conformational dynamics in delineating specificity.

THE HALOALKANOATE DEHALOGENASE SUPERFAMILY (HADSF)

The "central dogma" of protein structure/function studies is that protein sequence dictates protein structure which, in turn, defines protein function. Thus, evolution of any new function must be accompanied by the use of new sequence. In this way, as long as the "neutral drift" of sequences, allowing adoption of the same stable fold, can be distinguished from functionally transformative mutations, the prediction of function from sequence should be possible. In order to identify and assess sequence markers that support structure and specificity, we have undertaken the study of a large superfamily, comprised mostly of phosphoryltransferases, the haloalkanoate dehalogenase superfamily (HADSF) [1 – 3]. Because of the occurrence of the family in all domains of life and the number of homologues within each organism (27 in *E. coli* and 183 in *Homo sapiens* [4]) the members provide numerous examples of both orthologues to study determinants of specificity and paralogues to study function diversification.

The HADSF has successfully evolved several forms of chemical transformation and has experienced expansion through substrate space. Physiological substrates are varied in size and shape, ranging from phosphoproteins, nucleic acids, and phospholipids to phosphorylated disaccharides, sialic acids and terpenes to the smallest of the organophosphate metabolites, phosphoglycolate [5]. The diversity of substrates reflects the wide array of cellular roles of HADSF members including catalysis in biosynthetic pathways, reduction of antimetabolite levels, balancing nucleotide pools, elimination of toxins, and repair of damaged nucleic acids. The promiscuity and catalytic efficiency of the family matches the cellular function, with K_m values of $5 - 5000\,\mu\text{M}$ and with efficiency reflected by k_{cat}/K_m values ranging from $1 \times 10^3 - 1 \times 10^7\ \text{M}^{-1}\text{s}^{-1}$ (e. g. [6 – 9]).

Based on the work of Wolfenden [10], it is known that transfer of a phosphoryl group from a phosphate ester is demanding, such that the catalytic proficiency of the average HADSF member is 10^{17}. From this perspective, the catalytic machinery of the HADSF is well tuned to perform phosphoryl transfer. It is through the acquisition of structural appendages to the catalytic domain that the HADSF has succeeded in acting upon a vast array of substrates. The majority of the HADSF phosphatases consist of a conserved Rossmann domain (the catalytic domain) and a tethered "cap" domain that is variable in fold [3]. The catalytic scaffold is formed by four backbone segments located at the C-terminal end of the Rossmann-fold central sheet. The cap domain is inserted into either one of two loops of the catalytic scaffold (Figure 1).

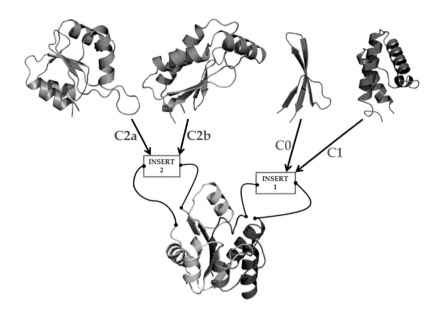

Figure 1. Cap assemblages of the HADSF can be broken down by insertion into the Rossman fold (shown as ribbon color ramped from blue to red) at site 1 of loops (type C 0) or alpha helical domains (type C 1), or at site 2 of mixed α/β domains (type C 2a and C 2b).

The catalytic scaffold binds the transferring phosphoryl group, whereas the cap domain binds the leaving group. Thus, HADSF phosphatase catalytic residues positioned by the catalytic scaffold are physically separated from the substrate recognition residues positioned on the cap domain. This trait is hypothesized to play a key factor in the evolution of the HADSF because the replacement of the substrate recognition residues that would switch the specificity from one physiological substrate to another would not perturb the environment of the catalytic residues, and hence their ability to stabilize the high energy transition states of the reaction pathway [4]. When the HADSF evolves new biochemical function by a switch of substrate there is no need for retooling the machinery responsible for catalysis. The beauty of this design is that one transition state fits all. Ultra-high resolution X-ray structures of a hexose phosphate phosphatase, BT4131 bound to transition-state analogues show that the catalytic domain residues in the HADSF form a "mold" in which the trigonal-bipyramidal transition states created during phosphoryl transfer are stabilized by electrostatic forces [11]. A conserved catalytic scaffold is used by which all members of the HADSF stabilize this transition-state geometry (Figure 2).

Allen, K.N. and Dunaway-Mariano, D.

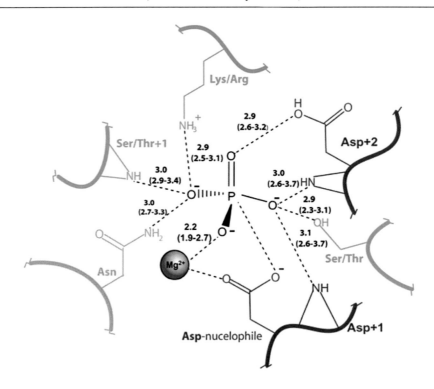

Figure 2. Composite scheme of the residues involved in forming the pentavalent trigonal–bipyramidal mold in the HADSF. Hydrogen bonds are depicted as dashed lines labeled with average bond lengths in Å (range of bond lengths in parenthesis). The notation +1 denotes the next amino acid in sequence. Loops/sequence motifs forming the active site from the Rossman fold are each colored differently (loop 1, red; loop 2, green; loop 3, blue; loop 4, red).

In concert with the residues contributed by the cap domain, dynamics between the catalytic domain and cap domain confer substrate specificity. Cap closure affords formation of the active site, which lays at the interface of the cap and the core domains. Ligand binding favours the cap closed form of the enzyme [12]. A classic example is that of β-phosphoglucomutase from *Lactoccocus lactis* (β-PGM), in which structural and kinetic analysis together have provided a model for the synchronization of cap closure with acid/base catalysis allowing the enzyme to discriminate between β-glucose 1-phosphate and water as the phosphoryl group acceptor [13]. Indeed mutation of one of the residues linking cap and core in β-PGM (Thr16) uncouples substrate binding from the positioning of the general acid/base, and changes the enzyme from a phosphomutase to a phosphohydrolase (the preference for hexose sugar over water as a phosphoryl acceptor changes from 6000:1 to 1:8 [14]).

Recognizing this intimate relationship between cap structure and enzyme function we have undertaken the function assignment of unknown members of the HADSF using an integrated approach through substrate profiling, structure analysis, computational substrate docking, and bioinformatic analysis [15]. This work highlights the promise and challenges ahead in function assignment by uncovering cases where: 1) structure and kinetics go hand in hand to assign activity and ultimately function, 2) structure and kinetics gives few clues to function and 3) subtleties of isoforms and allosteric effectors make assignment challenging.

STRUCTURE AND FUNCTION: HAND-IN-HAND

The HADSF member from *Bacteroides thetaiotaomicron*, BT2127, could easily be mistaken for an orthologue of β-PGM (sequence identity 22%, similarity 42%) because of the similarity in overall fold but also from the conserved His-Lys diad and linker hinge residue Ser/Thr that is characteristic of β-PGM (Figure 3). Surprisingly, *in vitro* substrate activity screening shows that BT2127 is not a mutase nor an organophosphate hydrolase but rather it is an inorganic pyrophosphatase with a k_{cat}/K_m value for pyrophosphate of ~1 × 10^5 M^{-1} s^{-1} [16]. The *in vitro* kinetic data, together with the gene context, support the assignment of *in vivo* function as an inorganic pyrophosphatase. Orthologues of β-PGM occur in species with maltose phosphorylase or trehalose phosphorylase (which collaborate in the utilization of maltose or trehalose as a carbon and energy source) and BT2127 occurs in species with bifunctional L-fucose kinase/L-fucose-1-phosphate guanyltransferase which produces inorganic pyrophosphate and β-L-fucose-GDP (used in biosynthesis of surface capsular polysaccharides and glycoproteins). The X-ray structural analysis of wild-type BT2127 shows that BT2127 differs from β-PGM in that the cap does not take on the open conformation in the absence of substrate. Specifically, Glu47 from the cap domain makes a coordinate bond to the Mg^{2+} ion in the core domain, promoting the closed conformer [16]. Thus, substrate discrimination is based, in part, on active-site size restrictions imposed by the cap domain. In BT2127 the substrate range and structure/function relationship are clear, but this is not always the case in the HADSF.

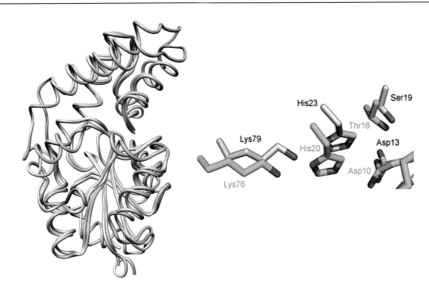

Figure 3. Superposition of BT2127 (PDB entry 3QX7) (gray) and *L. lactis* β-PGM (PDB entry 1O08) (cyan) showing the overall fold (left panel) and the conserved active-site residues (right panel).

GET A CLUE: STRUCTURE AND KINETICS MAY NOT YIELD CLUES TO FUNCTION

A hexose phosphate phosphatase in the HADSF from *Bacteroides thetaiotaomicron*, BT4131, possesses broad substrate specificities and low catalytic efficiency [8]. Within *Bacteroides* a homologue, BT1666, shares 38.4% sequence identity with BT4131. We queried whether these enzymes represent duplicated versus paralogous activities. The X-ray crystal structure of BT1666 compared to that of BT4131 reveals a conserved fold and identical active sites, suggestive of a common physiological substrate [17]. However, this apparent structural similarity is not mirrored in the substrate specificity profiles of the two enzymes using a panel of common phosphorylated metabolites (Table 1). BT1666, like BT4131, is an enzyme which shows "substrate blurring" or similar activities for a number of structurally similar substrates; however, the substrate specificity profiles for the two enzymes are distinct. Moreover, BT1666 shows overall lower activity (by approximately one order of magnitude) toward most substrates. This finding is unexpected as the active sites are identical and patterns of sequence conservation extremely similar. We posit that sequence variation outside the active site causes differences in conformational dynamics or subtle positioning effects that alter catalysis and thus drive the divergence in efficiency and selectivity. *In vivo* the overlapping substrate profiles may be explained by differential regulation of expression of the two enzymes or may confer an advantage in housekeeping functions by having a larger range of possible metabolites as substrates. There exists, of course, the possibility that the correct, specific substrate has not been uncovered for either enzyme. As seen in another example from the HADSF, subtleties of specificity can be introduced by effectors in the cellular milieu.

Table 1. Ratio of steady-state kinetic constants for BT4131 and BT1666 Catalyzed hydrolysis of phosphorylated small-molecule substrates in 50 mM HEPES containing 5 mM $MgCl_2$ (pH 7.0, 37 °C)

Substrate	BT1666/BT4131		
	k_{cat}/k_{cat}	K_m/K_m	$k_{cat}/K_m/k_{cat}/K_m$
D-glucose 6-phosphate	0.08	1.1	0.072
2-deoxy glucose 6-phosphate	0.02	3.3	0.006
glucosamine 6-phosphate	4.4	0.75	5.3
N-acetyl-glucosamine 6-phosphate	0.036	0.065	0.57
mannose 6-phosphate	0.23	1.3	0.18
fructose 6-phosphate	0.37	2	0.18
fructose 1,6-(bis)phosphate	78	0.62	125
arabinose 5-phosphate	0.027	0.43	0.063
ribose 5-phosphate	0.048	0.43	0.11
sorbitol 6-phosphate	0.078	0.96	0.081
DL-α-glycerol 3-phosphate	0.040	0.61	0.066
sucrose 6'-phosphate	8.5	0.45	19
trehalose 6-phosphate	5	–	–
ADP	28	6.3	4.5
pyridoxal 5'-phosphate	0.16	4.7	0.035
p-nitrophenyl-phosphate	0.66	0.66	1

FUNCTIONAL CHANGES EFFECTED BY LIGANDS

Human α-phosphomannomutase (α-PMM) catalyses the conversion of D-mannose 6-phosphate to α-D-mannose 1-phosphate which is required for GDP-mannose and dolichol-phosphate-mannose biosynthesis. These are essential constituents of N-glycosylation and glycosyl phosphatidylinositol membrane anchoring. Two isoforms PPM1 and PMM2 are found (65% sequence identity) with differential expression profiles (PMM2 is ubiquitous where PMM1 is expressed in brain and lungs) [18]. As might be supposed from their high sequence identity the two isozymes have a conserved fold and active-site structure (Figure 4) with root mean square deviations of 0.65 Å (for the cap) and 1.0 Å (for the core) when the two domains are overlaid separately [19].

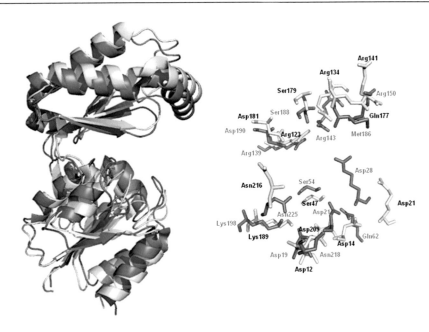

Figure 4. Superposition of PMM1 (PDB entry 2FUC) (gray) and PMM2 (PDB entry 2AMY; deposited in the Protein Data Bank by the Center for Eukaryotic Structural Genomics) (blue) showing the overall fold (left panel) and the active-site residues (right panel).

The isozymes are efficient catalysts of both substrates with high specificity as indicated by the magnitude of k_{cat}/K_m. Additionally, mutase activity requires that the bis-phosphorylated intermediates must dissociate from the active site (either within the enzyme or into bulk solvent) and then bind in the opposite orientation [20–22] so the bis-phosphorylated hexoses can also be considered substrates. Initially, it seems puzzling that such an essential enzyme would accept a second hexose, α-D-glucose 1-phosphate, as substrate. However, a classic observation brings to light a role for the glucose substrate. Two decades ago, a putative glucose-1,6-(bis)phosphatase brain enzyme was characterized [23–24] that is dependent for its activity on the presence of inosine mono-phosphate (IMP), the concentration of which increases in anoxia. Such an activity acts to restore glucose 6-phosphate levels and hence working ATP concentrations. More recently, the mystery phosphatase was identified as PMM1 [25]. Steady-state kinetics shows that IMP acts as an effector which shifts the activity of PMM1, but not PMM2, from that of a phosphomutases to a phosphohydrolase. Thus, in ischemic events, while PMM2 carries on the mutase activity to produce the α-mannose 1-phosphate essential to protein glycosylation, PMM1 switches to phosphohydrolase activity to restore glucose 6-phosphate levels for glycolysis. From a mechanistic perspective, we posit that IMP binds to the enzyme in place of the hexose phosphate, or to a secondary site to favour the conformation necessary for phosphoryl transfer, allowing water to act as the phosphoryl acceptor. This elegant solution to a metabolic problem exemplifies the challenges of using sequence/structure and even *in vitro* kinetics to predict *in vivo* function.

CONCLUSIONS

The explosion of sequences made available by genome sequencing efforts requires that current experimental strategies for function assignment must evolve into computationally-based function prediction. However, such a development must first find its basis in the expansion of the experimental database of sequence markers, notably, residues that support catalysis and specificity. Examination of the HADSF has allowed the analysis of a wide variety of orthologues and paralogues to develop the structure/function language. Remarkably, the analysis has revealed that while many enzymes are honed to perfection in terms of substrate selectivity, others show broad "substrate blurring" and even overlap of substrates between paralogues within a cell. This concept extends to isozymes where differential metabolic demands seem to play a part in retaining two very similar enzymes. In order to move from function in the test tube to role in the cell, auxiliary information such as regulation of transcription, cell environment, and presence and identity of regulating metabolites may be necessary.

ACKNOWLEDGEMENT

This work was supported by NIH U54 GM093342 to K.N.A. and D.D-M.

REFERENCES

[1]   Collet, J.F., van Schaftingen, E. Stroobant, V. (1998) A new family of phosphotrans-
        ferases related to P-type ATPases. *Trends Biochem. Sci.* **23**(8):284.
        doi: 10.1016/S0968-0004(98)01252-3.

[2]   Koonin, E.V. and Tatusov, R.L. (1994) Computer Analysis of Bacterial Haloacid
        Dehalogenases Defines a Large Superfamily of Hydrolases with Diverse
        Specificity: Application of an Iterative Approach to Database Search. *J. Mol. Biol.*
        244(1):125 – 132.
        doi: 10.1006/jmbi.1994.1711.

[3]   Burroughs, A.M., *et al.* (2006) Evolutionary genomics of the HAD superfamily:
        understanding the structural adaptations and catalytic diversity in a superfamily of
        phosphoesterases and allied enzymes. *J. Mol. Biol.* **361**(5): 1003 – 34.
        doi: 10.1016/j.jmb.2006.06.049.

[4]   Allen, K.N. and Dunaway-Mariano, D. (2009) Markers of fitness in a successful enzyme superfamily. *Curr. Opin. Struct. Biol.* **19**(6):658 – 65.
doi: 10.1016/j.sbi.2009.09.008.

[5]   Allen, K.N. and Dunaway-Mariano, D. (2004) Phosphoryl group transfer: evolution of a catalytic scaffold. *Trends Biochem. Sci.* **29**(9):495 – 503.
doi: 10.1016/j.tibs.2004.07.008.

[6]   Tremblay, L.W., Dunaway-Mariano, D. Allen, K.N. (2006) Structure and activity analyses of *Escherichia coli* K-12 NagD provide insight into the evolution of biochemical function in the haloalkanoic acid dehalogenase superfamily. *Biochemistry* **45**(4):1183 – 93.
doi: 10.1021/bi051842j.

[7]   Wang, L., *et al.* (2010) Divergence of biochemical function in the HAD superfamily: D-glycero-D-manno-heptose-1,7-bisphosphate phosphatase (GmhB). *Biochemistry* **49**(6):1072 – 81.
doi: 10.1021/bi902018y.

[8]   Lu, Z., Dunaway-Mariano, D. and Allen, K.N. (2005) HAD superfamily phosphotransferase substrate diversification: structure and function analysis of HAD subclass IIB sugar phosphatase BT4131. *Biochemistry* **44**(24):8684 – 96.
doi: 10.1021/bi050009j.

[9]   Wang, L., *et al.* (2008) Human symbiont Bacteroides thetaiotaomicron synthesizes 2-keto-3-deoxy-D-glycero-D-galacto-nononic acid (KDN). *Chem. Biol.* **15**(9):893 – 7.
doi: 10.1016/j.chembiol.2008.08.005.

[10]   Wolfenden, R. and Snider, M.J. (2001) The depth of chemical time and the power of enzymes as catalysts.*Acc. Chem. Res.* **34**(12):938 – 45.
doi: 10.1021/ar000058i.

[11]   Lu, Z., Dunaway-Mariano, D. Allen, K.N. (2008) The catalytic scaffold of the haloalkanoic acid dehalogenase enzyme superfamily acts as a mold for the trigonal bipyramidal transition state. *Proc. Natl. Acad. Sci. U.S.A.* **105**(15):5687 – 92.
doi: 10.1073/pnas.0710800105.

[12]   Zhang, G., *et al.* (2002) Kinetic Evidence for a Substrate-Induced Fit in Phosphonoacetaldehyde Hydrolase Catalysis. *Biochemistry* **41**(45):13370 – 13377.
doi: 10.1021/bi026388n.

[13]   Zhang, G., *et al.* (2005) Catalytic cycling in beta-phosphoglucomutase: a kinetic and structural analysis. *Biochemistry* **44**(27):9404 – 16.
doi: 10.1021/bi050558p.

[14] Dai, J., *et al.* (2009) Analysis of the structural determinants underlying discrimination between substrate and solvent in beta-phosphoglucomutase catalysis. *Biochemistry* **48**(9):1984 – 95.
doi: 10.1021/bi801653r.

[15] Gerlt, J.A., *et al.* (2001) The Enzyme Function Initiative. *Biochemistry* **50**(46):9950 – 62.
doi: 10.1021/bi201312u.

[16] Huang, H., *et al.* (2011) Divergence of structure and function in the haloacid dehalogenase enzyme superfamily: *Bacteroides thetaiotaomicron* BT2127 is an inorganic pyrophosphatase. *Biochemistry* **50**(41):8937 – 49.
doi: 10.1021/bi201181q.

[17] Lu, Z., Dunaway-Mariano, D. and Allen, K.N. (2011) The X-ray crystallographic structure and specificity profile of HAD superfamily phosphohydrolase BT1666: comparison of paralogous functions in *B. thetaiotaomicron*. *Proteins* **79**(11):3099 – 107.
doi: 10.1002/prot.23137.

[18] Pirard, M., *et al.* (1999) Kinetic properties and tissular distribution of mammalian phosphomannomutase isozymes. *Biochem. J.* **339**(1):201 – 207.
doi: 10.1042/0264-6021:3390201.

[19] Silvaggi, N.R., *et al.* (2006) The X-ray crystal structures of human alpha-phosphomannomutase 1 reveal the structural basis of congenital disorder of glycosylation type 1a. *J. Biol. Chem.* **281**(21):14918 – 26.
doi: 10.1074/jbc.M601505200.

[20] Dai, J., *et al.* (2006) Conformational cycling in beta-phosphoglucomutase catalysis: reorientation of the beta-D-glucose 1,6-(Bis)phosphate intermediate. *Biochemistry* **45**(25):7818 – 24.
doi: 10.1021/bi060136v.

[21] Rhyu, G.I., Ray, W.J., Jr. and Markley, J.L. (1984) Enzyme-bound intermediates in the conversion of glucose 1-phosphate to glucose 6-phosphate by phosphoglucomutase. Phosphorus NMR studies. *Biochemistry* **23**(2):252 – 60.
doi: 10.1021/bi00297a013.

[22] Naught, L.E. and Tipton, P.A. (2005) Formation and reorientation of glucose 1,6-bisphosphate in the PMM/PGM reaction: transient-state kinetic studies. *Biochemistry* **44**(18):6831 – 6.
doi: 10.1021/bi0501380.

[23] Guha, S.K. and Rose, Z.B. (1982) Brain glucose bisphosphatase requires inosine monophosphate. *J. Biol. Chem.* **257**(12):6634 – 7.

[24] Guha, S.K. and Rose, Z.B. (1983) Role of inosine 5'-phosphate in activating glu-
cose-bisphosphatase. *Biochemistry* **22**(6):1356 – 61.
doi: 10.1021/bi00275a006.

[25] Veiga-da-Cunha, M., *et al.* (2008) Mammalian phosphomannomutase PMM1 is the
brain IMP-sensitive glucose-1,6-bisphosphatase. *J. Biol. Chem.* **283**(49):33988 – 93.
doi: 10.1074/jbc.M805224200.

Experimental Standard Conditions of Enzyme Characterization
September 12th – 16th, 2011, Rüdesheim/Rhein, Germany

111

Functional (Mis)Assignment in the Tomaymycin Biosynthetic Pathway

Christian P. Whitman

Division of Medicinal Chemistry, College of Pharmacy,
The University of Texas at Austin, Austin, TX, 78712, U.S.A.

E-Mail: whitman@mail.utexas.edu

Received: 1st March 2012/Published: 15th February 2013

Abstract

4-Oxalocrotonate tautomerase (4-OT) catalyses the conversion of 2-hydroxymuconate to 2-oxo-3-hexenedioate in microbial pathways for the degradation of aromatic hydrocarbons. Pro-1 functions as a general base and shuttles the 2-hydroxy proton to C-5 of the product. Two arginine residues, Arg-11 and Arg-39, facilitate the reaction by participating in binding and catalysis. The same reaction is carried out by a heterohexamer 4-OT (hh4-OT) in thermophilic bacteria. The α-subunit of the hh4-OT identified the 4-OT homologue TomN in the biosynthetic cluster for the C ring of the antitumor antibiotic to-maymycin. TomN shares 58% pairwise sequence similarity with 4-OT including the three key catalytic residues. Kinetic and mutagenesis studies show that TomN catalyses the canonical 4-OT reaction with comparable efficiency using the same mechanism. However, the pro-posed function for TomN involves a very different reaction from that carried out by 4-OT. These results suggest that the assignment for TomN and the sequence of events leading to the C ring of tomaymycin might not be correct.

Introduction

Many bacterial species use aromatic compounds as their sole sources of carbon and energy because they have pathways to convert these compounds into substrates for the Krebs cycle (e.g., pyruvate and acetyl CoA) [1]. Initially, the aromatic compound is converted to catechol or a derivative. Subsequently, these catechols are processed by one of the many so-called meta-fission pathways. Meta-fission refers to a mode of catechol ring fission, as shown in Scheme 1. In the catechol meta-fission pathway, catechol 2,3-dioxygenase is responsible for ring-cleavage of catechol (**1**) to yield 2-hydroxymuconate semialdehyde (**2**). The aldehyde is oxidized by the NAD^+-dependent 2-hydroxymuconate semialdehyde dehydrogenase (2-HMSD), to yield 2-hydroxymuconate (**3**). Ketonisation of **3** to 2-oxo-3-hexenedioate (**4**) is catalysed by 4-oxalocrotonate tautomerase (4-OT). Decarboxylation of **4** by the metal-dependent 4-oxalocrotonate decarboxylase (4-OD) generates 2-hydroxy-2,4-pentadienoate (**5**). The metal-dependent vinylpyruvate hydratase (VPH) converts **5** to 4S-hydroxy-2-keto-pentanoate (**6**) by the addition of water. Cleavage of **6** by a pyruvate aldolase (PA) yields pyruvate and acetaldehyde. PA is tightly associated with acetaldehyde dehydrogenase, and this complex uses NAD^+ and CoASH to produce acetyl CoA from acetaldehyde [2].

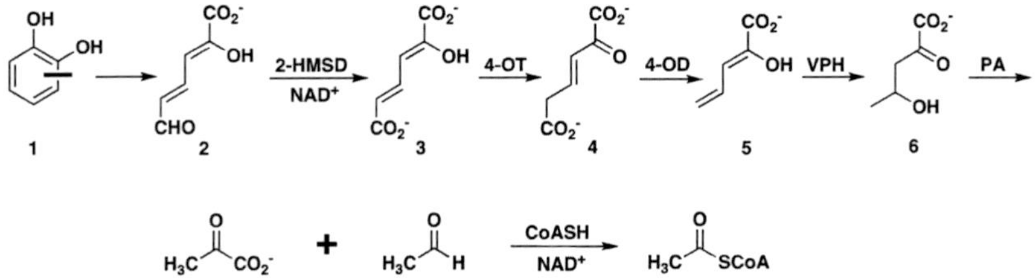

Scheme 1

The focus of this article will be the reactions catalysed by 4-oxalocrotonate tautomerase (4-OT) [3], the heterohexamer 4-oxalocrotonate tautomerase (hh4-OT) [4], and a 4-OT homologue known as TomN [5]. 4-OT was initially cloned from the TOL plasmid pWW0 in *Pseudomonas putida* mt-2. Organisms harbouring this plasmid can process simple aromatic hydrocarbons such as benzene, toluene, *m*- and *p*-xylene, 3-ethyltoluene, and 1,2,4-trimethylbenzene. The hh4-OT is found in the thermophilic organism *Chloroflexus aurantiacus* J-10-fl. The genomic context suggests that the hh4-OT is also part of a pathway for the degradation of aromatic hydrocarbons. TomN is found in the biosynthetic pathway for tomaymycin, a potent antitumor antibiotic agent.

The three enzymes are in the 4-OT family, which is one family in the tautomerase superfamily [6]. This superfamily is a group of structurally homologous proteins characterized by a β-α-β building block (see Figure 1) and a catalytic amino-terminal proline (Pro-1). Superfamily members are made up of short monomers (61 – 84 amino acids) or longer monomers, which are about twice as long. A short monomer codes for a single β-α-β unit, whereas a longer monomer codes for two β-α-β units that are connected by a short linker. Tautomerase superfamily enzymes carry out tautomerisation, dehalogenation, hydration, and decarboxylation reactions [7]. Thus far, Pro-1 is critical for all of these activities. As tautomerase superfamily members, the three enzymes share mechanistic and structural similarities. They also show subtle, but telling differences. The similarities and differences have implications for the assignment of function and the evolution of new activities in the tautomerase superfamily.

4-OXALOCROTONATE TAUTOMERASE

The mechanism and structure of the canonical 4-OT-catalyzed reaction (**3** to **4**, Scheme 2) have been studied for more than 20 years. The enzyme carries out a simple proton transfer reaction without the assistance of co-factors. A catalytic basein 4-OT abstracts the proton from the 2-hydroxy group of 2-hydroxymuconate (**3**) and places it at C-5 of 2-oxo-3-hexenedioate (**4**), in a highly stereoselective manner [3]. 4-OT is a hexamer, where each subunit is made up of 62 amino acids. It is also a founding member of the tautomerase superfamily.

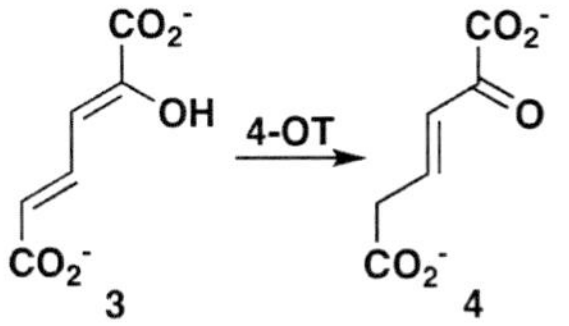

Scheme 2

Kinetic, mechanistic, and structural studies identified Pro-1, Arg-11, Arg-39, and Phe-50 (from different monomers) as key players in the 4-OT-catalyzed conversion of **3** to **4**. The proposed mechanism for 4-OT is shown in Scheme 3, where the primed residues refer to the different monomers. Pro-1, which has a pK_a of ~6.4 (determined by ^{15}N NMR titration), is the catalytic base responsible for the proton transfer from the 2-hydroxy group of **3** to C-5 of **4** [8]. The interaction between Arg-11 and the C-6 carboxylate group (of **3**) binds substrate and draws electron density to C-5 to facilitate protonation at C-5 [9]. Arg-39 interacts with the 2-hydroxy group and a carboxylate oxygen of C-1. Mutagenesis shows that the role of Arg-39 is primarily catalytic, where the positively charged guanidinium moiety stabilizes the developing carbanionic character after deprotonation of the 2-hydroxy group. Phe-50 (not shown) is a major contributor to a hydrophobic pocket near the prolyl nitrogen of Pro-1 [10]. The proximity of the pocket is largely responsible for the low pK_a of Pro-1.

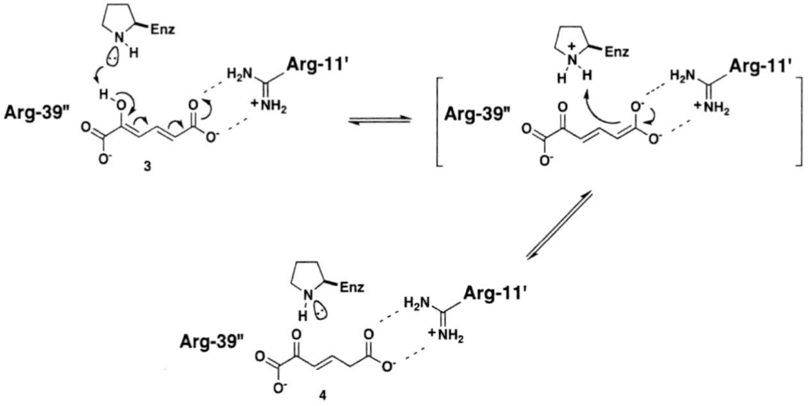

Scheme 3

The proposed mechanism for 4-OT is supported in part by the results of mutagenesis studies. The steady state kinetic parameters for the wild type enzyme and the P1A, R11A, and R39A mutants are summarized in Table 1. The first notable observation is that the wild type-catalysed reaction is near the diffusion-controlled limit. The second notable observation is the change in these kinetic parameters upon mutagenesis of the key residues. Changing Pro-1 to an alanine reduces k_{cat} (58-fold), but does not affect K_m [11]. The P1A mutant retains a catalytic base in the form of a primary amine so activity is not unexpected. Changing Arg-11 to an alanine has a major effect on K_m (9-fold increase) as well as a major effect on k_{cat} (87.5-fold reduction). These observations are consistent with the proposed binding and catalytic role. Finally, changing Arg-39 to an alanine has little effect on K_m, but a major effect on k_{cat} (125-fold reduction). These observations are consistent with a catalytic role for Arg-39.

Table 1. Kinetic parameters for 4-OT and mutants.

Enzyme	K_m (µM)	k_{cat} (s^{-1})	k_{cat}/K_m (M^{-1} s^{-1})
4-OT[a]	180 ± 30	3500 ± 500	1.9×10^7
P1A[b]	100 ± 12	60 ± 3	6.0×10^5
R11A[a]	1600 ± 300	40 ± 6	2.5×10^4
R39A[a]	290 ± 40	28 ± 2	9.7×10^4

[a]The steady state kinetic parameters were determined as described [9]. [b]The steady state kinetic parameters were determined as described elsewhere [11].

As noted above, 4-OT is a homohexamer. It is made up by the oligomerisation of a single β-α-β-building block (Figure 1) [12]. Two monomers form a dimer, and the three dimers assemble to form the hexamer (Figure 1). The GIGG motif (Gly-51, Ile-52, Gly-53, and

Glu-54) is critical for oligomer assembly and is a tell-tale sign of a hexamer in a short sequence. This motif is responsible for a β-hairpin at the end of each monomer, which interacts with the adjoining dimer and stabilizes the structure.

Figure 1. The 4-OT monomer (left), the 4-OT homohexamer (center), and a close-up of one of six active sites with the key residues labeled (right) (PDB code 4OTA). The 4-OT monomer shows signature tautomerase superfamily β–α–β building block. The catalytic amino-terminal proline is shown in space-filling form. The primed residues refer to different monomers.

There are six active sites per hexamer, which are located at the dimer interfaces (two per interface). A close-up of one active site shows the positions of the critical residues, Pro-1, Arg-11, Arg-39, and Phe-50 (Figure 1). These residues are contributed from three different monomers. Hence, activity is only observed for the hexamer, and not the individual dimers.

In the tautomerase superfamily, Pro-1 can function as a catalytic base or acid. In *trans*-3-chloroacrylic acid dehalogenase (CaaD), another 4-OT family member, Pro-1 functions as a catalytic acid with a pK_a of ~9.2 (determined by ^{15}N NMR titration) [13]. CaaD is found in a catabolic pathway for the nematacide, 1,3-dichloropropene (**7**, Scheme 4). In three enzyme-catalysed steps, **7** is converted to *trans*-3-chloroacrylate (**8**). Hydrolytic dehalogenation of **8** by CaaD produces malonate semialdehyde (**9**). Decarboxylation of **9** produces acetaldehyde, which is presumably channelled into the Krebs cycle. CaaD is a heterohexamer consisting of three α-subunits (with 75 amino acids) and three β-subunits (with 70 amino acids) [14]. The enzyme carries out the hydrolytic dehalogenation without the assistance of co-factors.

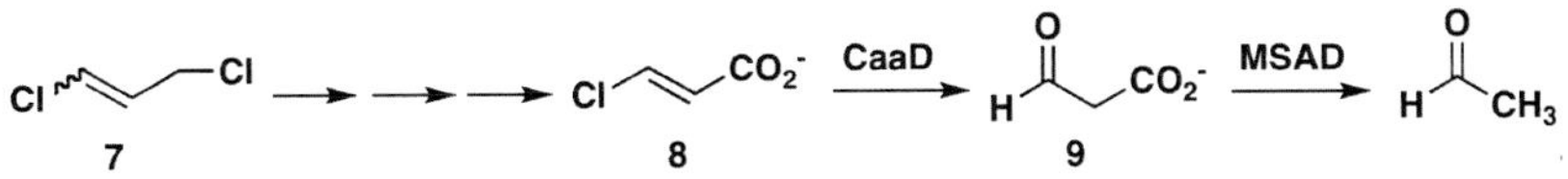

Scheme 4

The key players in the CaaD-catalysed conversion of **8** to **9** are βPro-1, αArg-8, αArg-11, and αGlu-52 (Scheme 5). αGlu-52 activates water for attack at C-3 of **8** (Scheme 5). The arginine residues, αArg-8 and αArg-11, interact with the C-1 carboxylate group, thereby binding and polarizing the substrate. The combined actions result in the enediolate intermediate **10.** The enediolate can undergo two fates. In path A, the enediolate picks up a proton from βPro-1 to generate the chlorohydrin intermediate **11.** Direct expulsion of the chloride produces **9.** In path B, the enediolate undergoes an α,β-elimination of HCl to yield **12.** Tautomerisation and protonation at C-2 by βPro-1 completes the reaction.

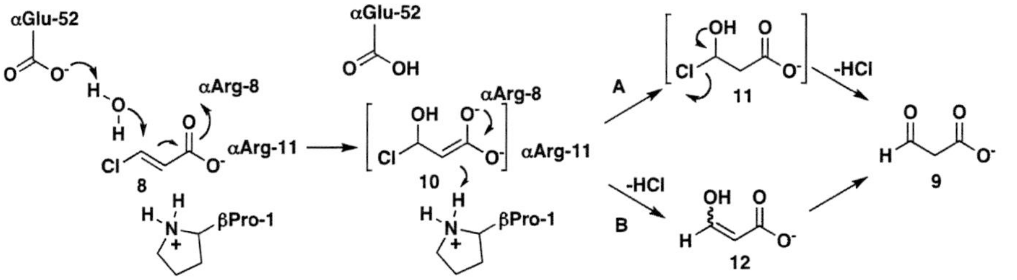

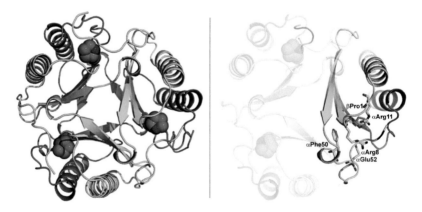

Scheme 5

The structure shows that an α- and β-subunit form a heterodimer (Figure 2) [14]. Three heterodimers form the CaaD heterohexamer. Although both subunits have an amino-terminal proline, only the βPro-1 is involved in catalysis. Hence, there are three active sites per heterohexamer. A close-up of one active site shows the positions of the critical residues (Figure 2). Notably, three key residues in CaaD (βPro-1, αArg-11, and αPhe-50) super-impose with those of 4-OT. A fourth residue (αArg-8) superimposes with Leu-8 in 4-OT.

Figure 2. The CaaD heterodimer in the heterohexamer (left) and a close-up of one of three active sites with the key residues labeled (right) (PDB code 3EJ3). The catalytic Pro-1 is shown in space-filling form.

These structural similarities coupled with the observation that both substrates (i. e., **3** and **8**) have an acrylate moiety suggested that 4-OT might have a low level CaaD activity. Subsequent experimentation confirmed that 4-OT does indeed have a low level CaaD activity, supporting the evolutionary link between CaaD and 4-OT [15]. The low level activity is dependent on Pro-1 and one of the two arginines (Arg-11 or Arg-39). One arginine might interact with the C-1 carboxylate group of **8** and create a partial positive charge at C-3

(Scheme 6). It is less clear what Pro-1 does in the reaction. The low level CaaD activity of 4-OT is a very clear example of catalytic promiscuity in the tautomerase superfamily and lends support to the idea that 4-OT and 4-OT homologues can serve as templates for the creation of new enzymatic activities (such as CaaD).

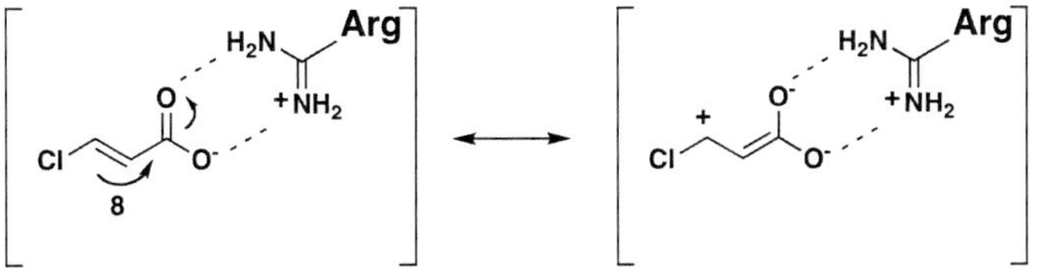

Scheme 6

The CaaD activity of 4-OT is enhanced in the L8R-4-OT [16]. This mutation increases the CaaD activity 50-fold (as assessed by the k_{cat}/K_m values), due mostly to a ~9-fold increase in k_{cat}. The mutation does not change the pK_a of Pro-1 or have structural consequences. The additional arginine might narrow the binding modes such that **8** is more likely to bind in catalytically productive mode. An attempt to further enhance the CaaD activity in 4-OT by adding an αGlu-52 equivalent was not successful. Ile-52 in 4-OT occupies the same position as αGlu-52 in CaaD. However, changing this isoleucine to a glutamate only reduced the 4-OT activity with no increase in CaaD activity.

A comparison of the k_{cat} value for the low level CaaD activity of 4-OT to that of CaaD does not suggest a robust reaction (Table 2) [16]. However, a comparison of this rate with that of the uncatalysed reaction shows that 4-OT enhances the reaction 10^8-fold (compared with the 10^{12}-fold rate enhancement exhibited by CaaD) [17]. The presence of the second arginine residue in the L8R-4-OT enhances the reaction 10^9-fold – a 10-fold increase. The L8R-4-OT only requires an additional 1000-fold rate enhancement to be a fully active CaaD. Pro-1, Arg-11 or Arg-39, and Arg-8 are determinants of the low level CaaD activity. One additional determinant was identified in the hh4-OT studies. However, mutations that might provide the 1000-fold rate enhancement have not been determined.

Table 2. The low-level CaaD activity of 4-OT.

Enzyme	k_{cat} (s^{-1})	Rate Enhancement
4-OT[a]	8.3×10^{-4}	~10^8
CaaD	3.8	~10^{12}
L8R-4-OT	8.8×10^{-3}	~10^9
No Enzyme[b]	2.2×10^{-12}	-

[a]The steady state kinetic parameters were determined as described elsewhere [15,16].
[b]The non-enzymatic rate was determined by Horvat and Wolfenden [17].

THE HETEROHEXAMER 4-OXALOCROTONATE TAUTOMERASE

In the course of ongoing sequence analysis, two 4-OTs were identified in the thermophile *Chloroflexus aurantiacus* J-10-fl. in a putative meta-fission pathway (based on sequence similarities with other meta-fission pathway enzymes) [4]. One 4-OT, with 72 amino acids, triggers the tautomerase annotation (in PSI-BLAST), but lacks Pro-1. A nearby 4-OT, also made up of 72 amino acids, has Pro-1, but does not trigger the tautomerase annotation. Both sequences have the two potentially catalytic arginines. A tryptophan residue is found in place of Phe-50 in the sequence that triggers the tautomerase annotation. At first, it was puzzling to find two 4-OTs in a meta-fission pathway because there is no apparent reason for two tautomerases. It also had potential implications for the tautomerase superfamily because it appeared that a member without a Pro-1 might have been discovered. Unfortunately, all attempts to express the two genes separately resulted in insoluble protein.

The mystery surrounding these two sequences was solved when co-expression of the two genes produced a stable heterohexamer where each dimer consists of an α- and β-subunit (Figure 3). The two subunits are arbitrarily designated the α- and β-subunits by their positions in the genome – the α-subunit appears first. The α-subunit sequence lacks Pro-1, but triggers the tautomerase annotation.

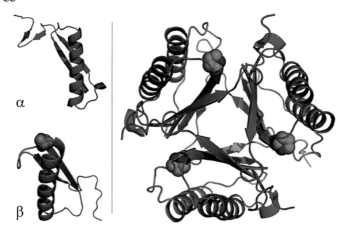

Figure 3. The α- and β-subunits of the hh4-OT (left) and the hh4-OT heterohexamer (right) (PDB code 3MB2). The three active sites are located at the interfaces. The catalytic amino-terminal proline is shown in space-filling form.

Both subunits have two arginine residues that could function in the mechanism. The α-subunit has Arg-12 and Arg-40 and the β-subunit has Arg-11 and Arg-39. Replacing the arginine residues in the β-subunit (with alanines) has only a minimal effect on the activity. However, changing βPro-1, αArg-12, or αArg-40 to an alanine has a major impact on the kinetic parameters for the reactions using **3** (Table 3) [4]. Moreover, the effects on the kinetic parameters parallel those seen with 4-OT. Changing βPro-1 to an alanine reduces k_{cat}

(83-fold) and K_m (4-fold). As noted for 4-OT, the P1A mutant has a catalytic base in the form of a primary amine so activity is expected. Changing αArg-12 to an alanine has significant effects on K_m (15-fold increase) and k_{cat} (70-fold decrease). These observations are consistent with the proposed binding and catalytic role of αArg-12. Finally, changing αArg-40 to an alanine increases K_m (5-fold) and reduces k_{cat} (46-fold). These observations are consistent with a catalytic role for αArg-40. These results identified βPro-1, αArg-12, and αArg-40 as critical residues, and indicate that the hh4-OT uses a mechanism like that proposed for 4-OT (see Scheme 3).

Table 3. Kinetic parameters for the hh4-OT and mutants[a].

Enzyme	K_m (μM)	k_{cat} (s^{-1})	k_{cat}/K_m (M^{-1} s^{-1})
4-OT	70 ± 8	3000 ± 100	4.3×10^7
βP1A	17 ± 2	36 ± 1	2.1×10^6
αR12A	1033 ± 510	43 ± 19	4.2×10^4
αR40A	345 ± 100	65 ± 14	1.9×10^5

[a]The steady state kinetic parameters were determined as described [4].

There is one notable difference between the hh4-OT and 4-OT – the hh4-OT lacks a low level CaaD activity. Incubation of **8** with a large quantity of the hh4-OT did not result in any detectable turnover after 7 weeks. This observation is somewhat surprising because an overlay of the two active sites shows little difference other than the replacement of Phe-50 (in 4-OT) with αTrp-51 (in hh4-OT) (Figure 4) [4].

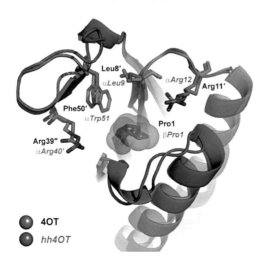

Figure 4. An overlay of the active sites of the hh4-OT and 4-OT. The active sites are similar except αTrp-51 in the hh4-OT replaces Phe-50 in 4-OT. The primed residues refer to different monomers.

However, the additional bulk on tryptophan might crowd Pro-1 and prevent alignment of **8** in the active site in such a way that it can undergo dehalogenation (Figure 5). This result identified Phe-50 as another determinant of the low level CaaD activity in 4-OT.

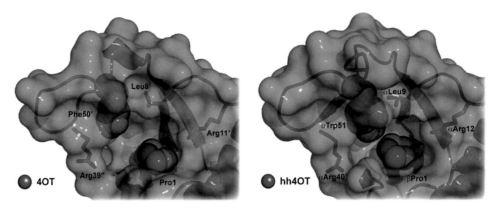

Figure 5. Space filling models showing the active sites of 4-OT (left) and the hh4-OT (right). αTrp-51 in the hh4-OT appears to crowd βPro-1 and might prevent alignment of **8** in the active site. The primed residues refer to different monomers.

TomN, a 4-Oxalocrotonate Tautomerase

The α-subunit of the hh4-OT led to the discovery of TomN [5]. TomN catalyses one step in the proposed biosynthetic pathway for the C ring of tomaymycin (**13**, Scheme 7) [18]. Tomaymycin and related pyrrolo[1,4]benzodiazepines (PBDs) such as sibiromycin are antibiotic antitumor agents [19]. The antitumor activity of these compounds results from sequence-specific DNA alkylation. Interest in this activity spawned the design, synthesis, and characterization of many derivatives. The chemical lability of the imine bond (see arrow in **13**, Scheme 7) complicates synthetic efforts and prompted the recent cloning of the sibiromycin and tomaymycin biosynthetic clusters. Biochemical characterization of the individual enzymes making up these pathways and manipulation of the corresponding genes might expand the repertoire of PBD analogues to include synthetically inaccessible ones and make feasible semi-synthetic approaches.

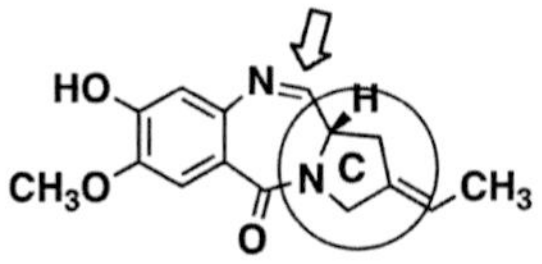

Tomaymycin (13)

Scheme 7

In the currently proposed pathway, hydroxylation of tyrosine produces L-dopa (14) (Scheme 8) [18]. TomH, a 2,3-dioxygenase, converts 14 to 16, presumably through 15 (as the *s-cis* or *s-trans* isomer) [20]. TomK processes 16 to 17 and oxalate. TomN catalyses the tautomerisation of 17 to 18. TomJ, an F420-dependent enzyme, carries out the reduction of the imine bond in 18 to yield 19, which is subsequently incorporated into 13. Although this scheme accounts for the construction of the C ring, the functions assigned to the individual enzymes are tentative and there is little biochemical evidence to support them.

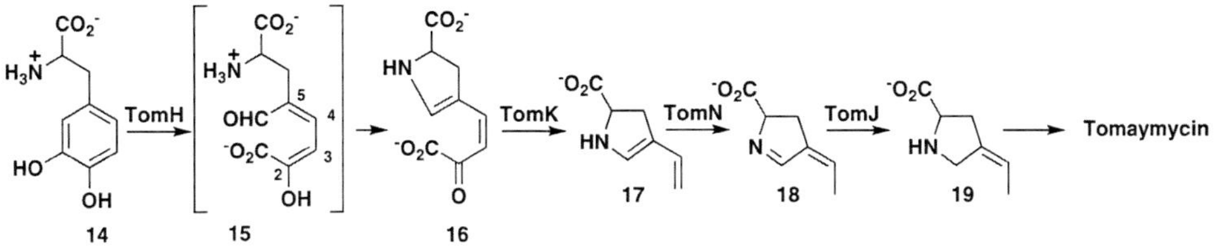

Scheme 8

The proposed TomN-catalysed reaction (17 to 18 in Scheme 8) is very different from the canonical 4-OT-catalyzed reaction (3 to 4 in Scheme 2). This observation is particularly striking because TomN shows 58% pairwise sequence similarity with 4-OT and has Pro-1, Arg-11, and Arg-39. In view of these similarities, it seemed odd that the reactions were so different and that TomN processes a monacid (i.e., 17), whereas 4-OT processes a diacid (i.e., 3). These observations prompted a kinetic and mutagenic analysis of TomN using 3 because the proposed TomN substrate is not available [5].

This analysis indicates that TomN carries out the canonical 4-OT reaction using the same residues (Pro-1, Arg-11, and Arg-39) with slightly less efficiency (5-fold, as judged by k_{cat}/K_m values) (Table 4). Changing Pro-1 to an alanine severely reduces k_{cat} (1233-fold), but has less of an effect on K_m (2.8-fold decrease). The change in k_{cat} is much more significant than that seen for the P1A mutant of 4-OT. Changing Arg-11 to an alanine results in a 2-fold increase in K_m and a 132-fold decrease in k_{cat}. The effect on K_m is not as significant as that measured for the R11A-4-OT, but is still consistent with a binding and catalytic role for Arg-11. Finally, changing Arg-39 to an alanine has little effect on K_m, but has a dramatic effect on k_{cat} (9250-fold reduction). These observations are consistent with a catalytic role for Arg-39, but the reduction in k_{cat} is much more severe than that measured for the R39A-4-OT.

Table 4. Kinetic parameters for TomN and mutants[a].

Enzyme	K_m (µM)	k_{cat} (s^{-1})	k_{cat}/K_m (M^{-1} s^{-1})
TomN	512 ± 225	1850 ± 630	3.6×10^6
P1A	180 ± 10	1.5 ± 0.1	8.3×10^3
R11A	1050 ± 240	14 ± 3	1.3×10^4
R39A	440 ± 60	0.2 ± 0.02	4.5×10^2

[a]The steady state kinetic parameters were determined as described [5].

The crystal structure for TomN shows that the enzyme is a hexamer constructed from the signature β-α-β building block (Figure 6).

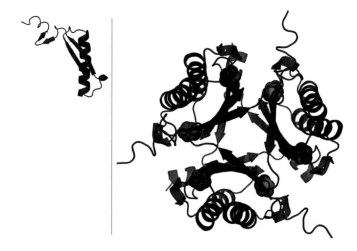

Figure 6. The TomN monomer (left) and the TomN homohexamer (right) (PDB code 3ry0). The catalytic amino-terminal proline is shown in space-filling form in the homohexamer.

The active site of TomN is very similar to those of 4-OT and the hh4-OT (not shown) (Figure 7) [5]. The key catalytic residues (Pro-1, Arg-11, and Arg-39) are positionally conserved. Trp-50 (in TomN) is found in place of Phe-50. These structural observations are consistent with the results of the kinetic and mutagenesis studies, indicating that TomN functions much like the canonical 4-OT.

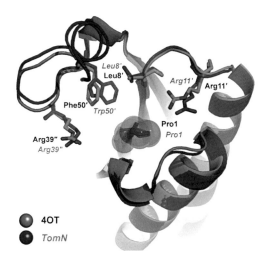

Figure 7. An overlay of the active sites of TomN and 4-OT. The active sites are similar except Trp-50 in the TomN replaces Phe-50 in 4-OT. The primed residues refer to different monomers.

The conversion of **3** to **4** is not likely the biological reaction for TomN, as suggested by the differences in K_m and k_{cat} for the wild type enzymes and the mutants. Nonetheless, the fact that TomN is an efficient 4-OT raises questions about the currently proposed function for TomN in the tomaymycin pathway. If TomN does not convert **17** to **18**, it might process a diacid substrate such as **15** (shown as the *s-trans* isomer in Scheme 9) or **20**, which is the cyclic counterpart of **15**. These possibilities are being examined. The outcome might also require a re-evaluation of the proposed sequence of events leading to the formation of the C ring and the functions of the enzymes.

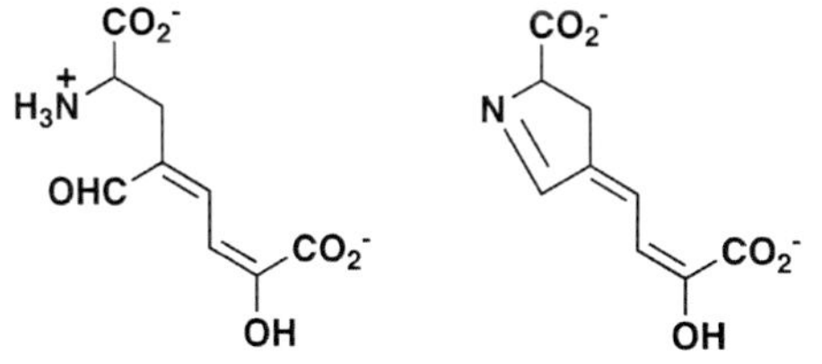

15 (s-trans)   **20**

Scheme 9

TomN is the first 4-OT (and the first tautomerase superfamily member) found in a biosynthetic pathway instead of a catabolic pathway. The group of biosynthetic 4-OTs is growing. 4-OT homologues have recently been identified in the pyridomycin and pristinamycin biosynthetic pathways (Pyr5 and SnbT, respectively). Pyridomycin is an anti-mycobacterial antibiotic with some unusual structural features [21]. Semi-synthetic water-soluble derivatives of pristinamycin I and II are the active constituents of Synercid, which is used for the treatment of various infections caused by Gram-positive bacteria and vancomycin-resistant *Enterococcus faecium* [22]. The roles of Pyr5 and SnbT in these pathways are not yet known, but elucidation of TomN's role in the tomaymycin pathway will provide insight.

CONCLUSIONS

Three 4-OTs (defined by their ability to convert **3** to **4**) have been characterized. They carry out the canonical 4-OT reaction with similar catalytic efficiencies and mechanisms, underscoring the difficulty of functional annotation for closely related family members in a superfamily. The assignment (or mis-assignment) of function for TomN best exemplifies the problem. Characterization of the enzymes has resulted in the discovery of a burgeoning group of biosynthetic 4-OTs.

ACKNOWLEDGMENTS

The research described in this paper was supported in part by the National Institutes of Health Grant GM-41239 and the Robert A. Welch Foundation Grant F-1334. The figures were generated by Dr. Gottfried K. Schroeder. In addition, the author is grateful to the numerous co-workers who contributed to the research described in this paper.

REFERENCES

[1]   Harayama, S., Rekik, M., Ngai, K.-L., and Ornston, L.N. (1989) Physically Associated Enzymes Produce and Metabolize 2-Hydroxy-2,4-dienoate, a Chemically Unstable Intermediate Formed in Catechol Metabolism via *meta* Cleavage in *Pseudomonas putida. J. Bacteriol.* **171**:6251–6258.

[2]   Manjasetty, B.A., Powlowski, J. and Vrielink, A. (2003) Crystal Structure of a Bifunctional Aldolase-dehydrogenase: Sequestering a Reactive and Volatile Intermediate. *Proc. Natl. Acad. Sci. U.S.A.* **100**:6992–6997.
doi: 10.1073/pnas.1236794100.

[3]   Whitman, C.P., Aird, B.A., Gillespie, W.R., and Stolowich, N.J. (1991) Chemical and Enzymatic Ketonization of 2-Hydroxymuconate, a Conjugated Enol. *J. Am. Chem. Soc.* **113**:3154–3162.
doi: 10.1021/ja00008a052.

[4]   Burks, E.A., Fleming, C.D., Mesecar, A.D., Whitman, C.P., and Pegan, S.D. (2010) Kinetic and Structural Characterization of a Heterohexamer 4-Oxalocrotonate Tautomerase from *Chloroflexus aurantiacus* J-10-fl: Implications for Functional and Structural Diversity in the Tautomerase Superfamily. *Biochemistry* **49**:5016 – 5027. doi: 10.1021/bi100502z.

[5]   Burks, E.A., Yan, W., Johnson, Jr., W.H., Li, W., Schroeder, G.K., Min, C., Gerratana, B., Zhang, Y., and Whitman, C.P. (2011) Kinetic, Crystallographic, and Mechanistic Characterization of TomN: Elucidation of a Function for a 4-Oxalocrotonate Tautomerase Homologue in the Tomaymycin Biosynthetic Pathway. *Biochemistry* **35**:7600 – 7611. doi: 10.1021/bi200947w.

[6]   Whitman, C.P. (2002) The 4-Oxalocrotonate Tautomerase Family of Enzymes: How Nature Makes New Enzymes Using a β-α-β Structural Motif. *Arch. Biochem. Biophys.* **402**:1 – 13. doi: 10.1016/S0003-9861(02)00052-8.

[7]   Poelarends, G.J., Veetil, V.P. and Whitman, C.P. (2008) The Chemical Versatility of the β-α-β Fold: Catalytic Promiscuity and Divergent Evolution in the Tautomerase Superfamily. *Cell. Mol. Life Sci.* **65**: 3606 – 3618. doi: 10.1007/s00018-008-8285-x.

[8]   Stivers, J.T., Abeygunawardana, C., Mildvan, A.S., Hajipour, G., and Whitman, C. P. (1996) 4-Oxalocrotonate Tautomerase: pH Dependences of Catalysis and pK_a Values of Active Site Residues. *Biochemistry* **35**:814 – 823. doi: 10.1021/bi9510789.

[9]   Harris, T.K., Czerwinski, R.M., Johnson, Jr., W.H., Legler, P.M., Abeygunawardana, C., Massiah, M.A., Stivers, J.T., Whitman, C.P., and Mildvan, A.S. (1999) Kinetic, Stereochemical, and Structural Effects of Mutations of the Active Site Arginine Residues in 4-Oxalocrotonate Tautomerase. *Biochemistry* **38**:12343 – 12357. doi: 10.1021/bi991116e.

[10]   Czerwinski, R. M., Harris, T.K., Massiah, M.A., Mildvan, A.S., and Whitman, C.P. (2001) The Structural Basis for the Perturbed pK_a of the Catalytic Base in 4-Oxalocrotonate Tautomerase: Kinetic and Structural Effects of Mutations of Phe-50. *Biochemistry* **40**:1984 – 1995. doi: 10.1021/bi0024714.

[11]   Czerwinski, R.M., Johnson Jr., W.H., Whitman, C.P., Harris, T.K., Abeygunawardana, C., and Mildvan, A.S. (1997) Kinetic and Structural Effects of Mutations of the Catalytic Amino-terminal Proline in 4-Oxalocrotonate Tautomerase. *Biochemistry* **36**:14551 – 14560. doi: 10.1021/bi971545 h.

[12] Subramanya, H.S., Roper, D.I., Dauter, Z., Dodson, E.J., Davies, G.J., Wilson, K.S., and Wigley, D.B. (1996) Enzymatic Ketonization of 2-Hydroxymuconate: Specificity and Mechanism Investigated by the Crystal Structures of Two Isomerases. *Biochemistry* **35**:792 – 802.
doi: 10.1021/bi951732k.

[13] Wang, S.C, Person, M.D., Johnson, Jr., W.H., and Whitman, C.P. (2003) Reactions of *trans*-3-Chloroacrylic Acid Dehalogenase with Acetylene Substrates: Consequences of and Evidence for a Hydration Reaction. *Biochemistry* **42**:8762 – 8773.
doi: 10.1021/bi034598+.

[14] de Jong, R.M., Brugman, W., Poelarends, G.J., Whitman, C.P., and Dijkstra, B.W. (2004) The X-ray Structure of *trans*-3-Chloroacrylic Acid Dehalogenase Reveals a Novel Hydration Mechanism in the Tautomerase Superfamily. *J. Biol. Chem.* **279**:11546 – 11552.
doi: 10.1074/jbc.M311966200.

[15] Wang, S.C., Johnson, Jr. W.H., and Whitman, C.P. (2003) The 4-Oxalocrotonate Tautomerase- and YwhB-catalyzed Hydration of 3*E*-Haloacrylates: Implications for Evolution of New Enzymatic Activities. *J. Am. Chem. Soc.* **125**:14282 – 14283.
doi: 10.1021/ja0370948.

[16] Poelarends, G.J., Almrud, J.J., Serrano, H., Darty, J.E., Johnson, Jr., W.H., Hackert, M.L., and Whitman, C.P. (2006) Evolution of Enzymatic Activity in the Tautomerase Superfamily: Mechanistic and Structural Consequences of the L 8R Mutation in 4-Oxalocrotonate Tautomerase. *Biochemistry* **45**:7700 – 7708.
doi: 10.1021/bi0600603.

[17] Horvat, C.M. and Wolfenden, R.V. (2005) A Persistent Pesticide Residue and the Unusual Catalytic Proficiency of a Dehalogenating Enzyme. *Proc. Natl. Acad. Sci. U.S.A.* **102**:16199 – 16202.
doi: 10.1073/pnas.0508176102.

[18] Li, W., Chou, S.C., Khullar, A., and Gerratana, B. (2009) Cloning and Characterization of the Biosynthetic Cluster for Tomaymycin, an SJG-136 Monomeric Analog. *Appl. Environ. Microbiol.* **75**:2958 – 2963.
doi: 10.1128/AEM.02325-08.

[19] Li, W., Khullar, A., Chou, S.C., Sacramo, A., and Gerratana, B. (2009) Biosynthesis of Sibiromycin, a Potent Antitumor Antibiotic. *Appl. Environ. Microbiol.* **75**:2869 – 2878.
doi: 10.1128/AEM.02326-08.

[20]   Colabroy, K.L., Hackett, W.T., Markham, A.J., Rosenberg, J., Cohen, D.E., and Jacobson, A. (2008) Biochemical Characterization of L-DOPA 2,3-Dioxygenase, a Single-Domain Type I Extradiol Dioxygenase from Lincomycin Biosynthesis. *Arch. Biochem. Biophys.* **479**:131 – 138.
doi: 10.1016/j.abb.2008.08.022.

[21]   Huang, T., Wang, Y., Yin, J., Du, Y., Tao, M., Xu, J., Chen, W., Lin, S., and Deng, Z. (2011) Identification and Characterization of the Pyridomycin Biosynthetic Gene Cluster of *Streptomyces pyridomyceticus* NRRL B-2517. *J. Biol. Chem.* **286**:20648 – 20657.
doi: 10.1074/jbc.M110.180000.

[22]   Mast, Y., Weber, T., Golz, M., Ort-Winklbauer, R., Gondran, A., Wohlleben, W., and Schinko, E. (2011) Characterization of the 'Pristinamycin Supercluster' of *Streptomyces pristinaespiralis*. *Microb.Biotechnol.* **4**:192 – 206.
doi: 10.1111/j.1751-7915.2010.00213.x.

THERMODYNAMIC CHARACTERISATION OF CARBOHYDRATE-ACTIVE ENZYMES

OLIVER EBENHÖH[1,*], ALEXANDER SKUPIN[2], ÖNDER KARTAL[3], SEBASTIAN MAHLOW[4] AND MARTIN STEUP[4]

[1]Institute for Complex Systems and Mathematical Biology, University of Aberdeen, Aberdeen, UK

[2]Luxembourg Centre for Systems Biomedicine, University of Luxembourg, Luxembourg

[3]Group of Plant Biotechnology, Department of Biology, ETH Zürich, Zürich, Switzerland

[4]Department of Plant Physiology, Institute of Biochemistry and Biology, University of Potsdam, Potsdam-Golm, Germany

E-MAIL: *oliver.ebenhoeh@googlemail.com

Received: 22ⁿᵈ March 2012/Published: 15ᵗʰ February 2013

ABSTRACT

Many carbohydrate-active enzymes catalyse a specific reaction pattern instead of one particular reaction. For example, glucanotransferases and glucosyltransferases recognise the reducing ends of glucans irrespective of their degree of polymerisation. Thus, in principle, they are capable of catalysing an infinite number of reactions. Here we show how concepts from statistical thermodynamics can be employed to characterise the action patterns of polymer-active enzymes and determine their equilibrium distributions. For selected enzymes, we provide experimental evidence that our theory provides accurate predictions. We further outline how the thermodynamic description can be employed to experimentally determine bond energies from equilibrium distributions of polymer-active enzymes.

INTRODUCTION

The classical thermodynamic characterization of enzymes is based on the thermodynamic equilibrium K_{eq}. However, this equilibrium constant is only defined if the enzyme under consideration catalyses exactly one reaction of the general form

$$\sum_i \nu_i A_i = 0, \tag{1}$$

in which A_i are the involved chemical species and the ν_i are the stoichiometric coefficients, denoting how many molecules are consumed (negative ν_i) or produced (positive ν_i) per reaction.

In carbohydrate metabolism many enzymes important in the synthesis and degradation of biopolymers do not follow this rule. A prominent example are glucanotransferases [1] specifically binding the non-reducing end of a glucan, however irrespective of its exact length, or degree of polymerisation (DP). This leads to the problem that glucanotransferases mediate an infinite number of theoretically possible transfer reactions. For example disproportionating enzyme 1 (DPE1), which plays an important role in the starch breakdown pathway by metabolising maltotriose, transfers $q = 1,2,3$ glucosyl residues from one glucan to another, resulting in the chemical reactions

$$G_n + G_m \rightleftharpoons G_{n-q} + G_{m+q}. \tag{2}$$

In this particular example, every reaction (2) is even energetically neutral [2], leading to $K_{eq} = 1$. Since the enzyme acts on a polydisperse mixture containing a large number of specific reactants, the characterisation of the thermodynamic equilibrium by the classical equilibrium constant is clearly insuffient.

An approach leading to a satisfying enzymatic characterisation is based on a description of polydisperse mixtures as statistical ensembles, which we recently proposed [3]. A glucan of DP n is described as a particle with energy state E_{n-1}, reflecting the energy stored in the interglucosidic bonds. In this framework, an enzyme such as DPE1 mediates a transition of particles between energy states, where the possible transitions reflect the enzymatic mechanism. For example, DPE1 will simultaneously shift one particle one, two or three energy levels up and one particle the same number of levels down.

In this paper, we will exploit this analogy and derive a consistent thermodynamic characterisation of the polydisperse equilibrium. We show how this understanding can be used to determine bond energies from experimentally observed equilibrium distributions.

THEORY

A mixture of dissolved substances can be explicitly characterised by its Gibbs energy which is conveniently written as

$$G = G^f - T\tilde{S}_{\text{mix}}. \tag{3}$$

Here, G^f is the total energy of formation and $\tilde{S}_{\text{mix}}$ is the mixing entropy. Numbering the dissolved substances by i and denoting their respective concentrations with c_i, the total energy of formation is

$$G^f = \sum_i c_i \Delta_f G_i^0, \tag{4}$$

where $\Delta_f G_i^0$ denotes the standard Gibbs energy of formation of substance i, and the mixing entropy of the solution can be written as [3, Supplementary information]

$$\tilde{S}_{\text{mix}} = -R \sum_i c_i (\ln c_i - 1). \tag{5}$$

Any reaction system acting on such a mixture of dissolved reactants changes the composition of the solution and therefore its formation energy and mixing entropy in such a way that the Gibbs energy decreases [4]. To identify the equilibrium, the minimum of the Gibbs energy has to be determined, taking into account that the concentrations of the dissolved substances cannot change completely arbitrarily but are constrained through conservation rules imposed by the reaction network. If $\mathbf{N}$ is the stoichiometric matrix of the reaction network, the conservation rules are defined by the left-sided kernel [5]

$$\mathbf{L} \cdot \mathbf{N} = 0. \tag{6}$$

There exist $k = \text{rank } \mathbf{L}$ independent conserved quantitites, which in their most general form can be written as

$$\sum_j l_{ij} c_j = b_i, \quad \text{with} \quad i = 1 \dots k, \tag{7}$$

where the values l_{ij} are stoichiometric coefficients and b_i are constants that are defined by the initial conditions and thus are experimentally controllable. The equilibrium of any reaction system can consequently be calculated by minimising the Gibbs energy (3) under the constraints (7).

Glucanotransferases mediate bimolecular reactions of the type (2), converting two substrate molecules in two product molecules by transferring a number of residues from one to the other. Consequently, a glucanotransferase in isolation catalyses a reaction network in which the number of molecules is conserved,

$$c^{\text{tot}} = \sum c_i = \text{const.}, \tag{8}$$

and it is convenient to introduce the molar fractions

$$x_i = \frac{c_i}{c^{\text{tot}}},$$ (9)

leading to

$$\sum c_i(\ln c_i - 1) = \sum c_i \ln c_i - c^{\text{tot}} = c^{\text{tot}}\left[\sum x_i \ln x_i + \ln c^{\text{tot}} - 1\right].$$ (10)

Since $\ln(c^{\text{tot}}) - 1$ is constant, it is irrelevant for the determination of the minimum of the Gibbs free energy. Defining the molar energies of formation by $G^f = c^{\text{tot}} g^f$ allows to define the molar Gibbs free energy

$$g = g^f - RT \sum x_i \ln x_i = g^f - TS,$$ (11)

where the mixing entropy

$$S = -R \sum x_i \ln x_i,$$ (12)

now assumes its well-known form. Notably, this definition is independent on the choice of the units in which concentrations are measured. In contrast, $\tilde{S}_{\text{mix}}$ as defined in Eq. (5) is dependent on the concentration units.

CHARACTERISATION OF THE EQUILIBRIUM OF VARIOUS TRANSFERASES

Disproportionating enzyme 1 (DPE1)

Disproportionating enzyme 1 (DPE1) catalyses reactions according to formula (2) mentioned in the Introduction, where $q = 1,2,3$ is the number of transferred glucosyl residues. Obviously, this reaction conserves the total number of molecules,

$$\sum x_i = 1.$$ (13)

This reaction is energetically neutral in the sense that the bond enthalpies of the cleaved α-1,4 glucosidic linkage and the formed α-1,4 glucosidic linkage are identical [2]. Thus, the catalytic activity of DPE1 does not change g^f and minimising g in Eq. (11) is equivalent to maximising

$$\frac{S}{R} = -\sum x_i \ln x_i.$$ (14)

Since the enzyme is only attacking glucosidic linkages between the glucosyl residues, we identify a glucan of DP n with an energy state E_{n-1} containing the bond enthalpies of $n - 1$ α-1,4 glucosidic linkages.

Thermodynamic characterisation of carbohydrate-active enzymes

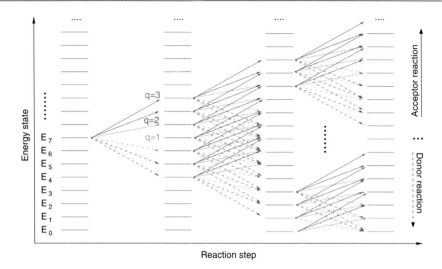

Figure 1. Scheme of the DPE1 mediated reaction system. DPE1 mediates transfers of glucose, maltose and maltotriose units, i.e. $q = 1,2,3$. In each reaction step the system follows an arbitrary dashed and solid arrow of the same colour simultaneously. This leads to a combinatorial explosion of the reaction system. The lower limit of DP leads to a reflecting boundary condition for G_1 which causes the Boltzmann distribution.

As depicted in Figure 1, the catalytic action of DPE1 corresponds to the simultaneous downward shift of one molecule from an energy state E_n to E_{n-q} (donor reaction) and an upward shift of another molecule from energy state E_m to E_{m+q} (acceptor reaction), where $q = 1,2,3$. We denote by x_i the molar fraction of molecule in energy state E_i. Thus, the molar fraction of glucose molecules is x_0 and that of a glucan of DP n is x_{n-1}.

Apart from the conservation of the total number of molecules, Eq. (13), the enzyme action also conserves the total number of interglucose bonds. Thus

$$\sum k \cdot x_k = b, \tag{15}$$

where the constant b describes the average number of bonds per glucan. This number is determined by the initial condition with which the reaction is initiated and is therefore an experimentally controllable parameter.

To obtain the molar fractions $\{x_i\}$ which maximise the entropy defined by Eq. (14), we define the Lagrange function

$$L(x_k; \alpha, \beta) = -\sum_k x_k \ln x_k - \alpha \left(\sum_k x_k - 1 \right) - \beta \left(\sum_k k \cdot x_k - b \right) \tag{16}$$

and set the partial derivatives to zero,

$$0 = \frac{\partial L}{\partial x_0} = -(\ln x_0 + 1) - \alpha, \tag{17}$$

$$0 = \frac{\partial L}{\partial x_k} = -(\ln x_k + 1) - \alpha - k \cdot \beta. \tag{18}$$

This yields

$$x_k = x_0 \cdot e^{-k\beta} = x_0 \, y^k, \tag{19}$$

where $y = e^{-\beta}$ is introduced for convenience. This result demonstrates that in equilibrium the degrees of polymerisation are exponentially distributed. The specific values for x_0 and y are determined from the constraints, where it is convenient to exploit the analogy to the formalism in statistical physics and introduce the partition function $Z = \sum y^k$, such that

$$x_0 = \frac{1}{Z} \quad \text{and} \quad b = \frac{y}{Z} \frac{\partial Z}{\partial y}. \tag{20}$$

These expressions fully characterise the equilibrium distribution. The entropy in equilibrium, S_{eq}, is given by

$$S_{eq}/R = -\sum_k x_0 y^k \ln(x_0 y^k) = -\ln x_0 - b \ln y = \ln Z - \ln y \cdot \frac{\partial \ln Z}{\partial \ln y}. \tag{21}$$

These expressions are valid regardless of the precise range over which the sums in Eqs. (13), (15) and (16) have to be extended. Because DPE1 recognises the non-reducing end of a glucan, there is no *a priori* reason to assume any limit for the DP. If the sums are extended over all numbers $k \in \{0, 1, 2, \ldots\}$, the resulting expressions have a particularly simple form. The partition sum reads

$$Z = \sum_{k=0}^{\infty} y^k = \frac{1}{1-y}. \tag{22}$$

It follows that

$$b = \frac{y}{1-y} \Leftrightarrow y = \frac{b}{b+1} \tag{23}$$

and

$$x_0 = \frac{1}{b+1}. \tag{24}$$

Thus, the equilibrium distribution is given by

$$x_k = (1 - y) y^k = \left(1 - e^{-\beta}\right) e^{-\beta k}. \tag{25}$$

We have experimentally tested our predictions for various initial degrees of polymerisation, defining the parameter b.

Thermodynamic characterisation of carbohydrate-active enzymes

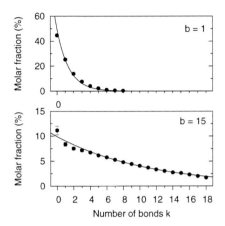

Figure 2. Predicted and measured equilibrium distributions for two different initial conditions. In the upper panel, the reaction was initiated with maltose, corresponding to $b = 1$. In the lower panel, the initial distribution had an average degree of polymerisation of 16, corresponding to $b = 15$. Solid lines indicate the predicted equilibrium distribution, symbols represent experimentally determined values. Error bars indicate the standard deviation determined from three independent measurements.

In Figure 2, predicted and observed equilibrium distributions are shown for two initial conditions. We have tested the case for extremely short DPs by incubating the reaction with maltose ($b = 1$, upper panel) and for long DPs by incubating the reaction with a mixture of long glucans ($b = 15$, lower panel).

Clearly, $0 \leq y \leq 1$ holds for all positive values of b, implying that β is always positive. In the limit of very long initial DPs,

$$\lim_{b \to \infty} y = 1. \tag{26}$$

The entropy in equilibrium equals

$$S_{eq}/R = - \sum_{k=0}^{\infty} x_k \ln x_k = (b+1)\ln(b+1) - b \ln b. \tag{27}$$

In Figure 3, experimentally determined values for the equilibrium entropy are depicted together with the theoretical predictions according to Eq. (27).

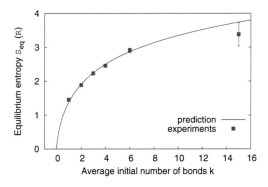

Figure 3. Predicted and measured values for the entropy in equilibrium for different initial conditions. The solid line represents the theoretical prediction of the maximal entropy and the squares represent experimentally determined equilibrium entropies. Error bars are standard deviations of three independent experiments.

Various experimental studies suggested that DPE1 cannot utilise maltose as glucosyl donor and cannot form maltose as product [6, 7, 8]. We could show [3] that this is only approximately true, but that in fact maltose serves as a poor substrate with an approximately 800-fold reduced binding affinity when compared to glucans of other DPs. It is, however, possible to analytically find the equilibrium under the assumption that maltose strictly does not participate in the DPE1-mediated reactions. In this case, maltose (x_1) has to be excluded from the sums and the partition sum is

$$Z = \sum_{\substack{k=0 \\ k \neq 1}}^{\infty} y^k = \frac{1-y+y^2}{1-y} = \frac{1}{x_0} \tag{28}$$

and relations (20) allow to find the implicit equation determining y from b,

$$b = \frac{y}{1-y} \cdot \frac{2y-y^2}{1-y+y^2} = \frac{y}{1-y} + \frac{2y^2-y}{1-y+y^2}. \tag{29}$$

The corresponding value of the equilibrium parameter β describes the quasi-equilibrium which is observed in experiments when the incubation time is long enough for the equilibration of all glucans except maltose but still too short for the detection of maltose. The effect of excluding maltose from the sums is depicted in Figure 4.

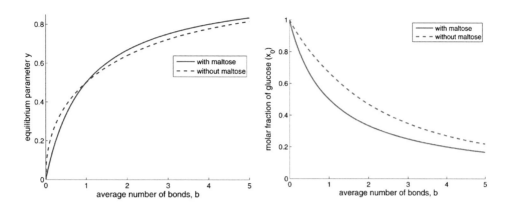

Figure 4. The equilibrium paramter y and the equilibrium concentration of glucose (x_0) in dependence on the initial conditions reflected by average number of bonds (b) for the case in which all DPs are included (solid lines) and the case in which maltose (x_1) is excluded from the sums (dashed lines).

Disproportionating enzyme 2 (DPE2)

DPE2 catalyses the transfer of a single glucose residue from one α-1,4-linked glucan to another. It therefore also belongs to the class of energetically neutral enzymes and obeys the constraints of conserved number of molecules (13) and conserved total number of bonds (15). However, as our and other [9] experimental findings suggest, maltose never acts as an acceptor of glucosyl residues and maltotriose never acts as a donor. DPE2 therefore catalyses reactions according to the formula

$$G_n + G_m \rightleftharpoons G_{n-1} + G_{m+1}, \quad \text{where } n \neq 3 \text{ and } m \neq 2. \tag{30}$$

Like for DPE1, we again associate a glucan with DP n with an energy state E_{n-1} and denote the corresponding molar fraction by x_{n-1}. Similar to DPE1, the enzymatic action of DPE2 corresponds to a simultaneous down-shift of one molecule from energy state E_n to E_{n-1} and an up-shift of another molecule from energy state E_m to E_{m+1}.

The additional enzymatic constraint that maltose cannot accept glucosyl residues and maltotriose cannot act as donor implies an additional constraint, namely the conservation of the sum of glucose and maltose molecules,

$$x_0 + x_1 = p, \tag{31}$$

where p is determined by the initially applied amount of glucose and maltose. The DPE2 mediated reaction scheme is shown in Figure 5 where the separation of the glucose-maltose pool from the pool of larger DPs is shown by the red dashed line which is not crossed by any possible reaction path. In each DPE2 reaction step, one arbitrary donor reaction (dashed arrows) occurs simultaneously with one arbitrary acceptor reaction (solid arrows). Starting from an initial substrate mixture of maltohexaose (E_5) and maltose (E_1), the 5 first possible reactions are shown in Figure 5, where in each step the reaction system follows a dashed and a solid line simultaneously.

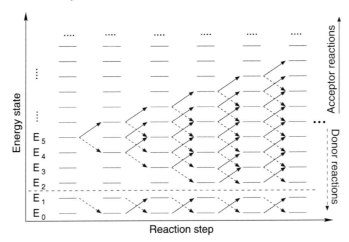

Figure 5. Scheme of the DPE2 mediated reaction system. Each DPE2 reaction step consists of one donor and one acceptor reaction depicted by a dashed and a solid arrow, respectively. Due to the restriction that maltose is never an acceptor and maltotriose is never a donor, the maltose and glucose pool is separated from the other DPs as shown by the red dashed line. The scheme exhibits all possible reaction pathways starting from the two indicated initial substrates maltohexaose and maltose, where in each step one arbitrary solid and one arbitrary dashed path is taken.

Again, the maximal entropy is determined using the method of Lagrangian multipliers. The Lagrangian

$$L(x_k; \alpha, \beta, \gamma) = -\sum_k x_k \ln x_k - \alpha \left(\sum_k x_k - 1 \right) - \beta \left(\sum_k k \cdot x_k - b \right) - \gamma(x_0 + x_1 - p) \quad (32)$$

now contains an additional Lagrange multiplier γ reflecting the new contraint (31). Setting the partial derivatives to zero,

$$0 = \frac{\partial L}{\partial x_0} = -(\ln x_0 + 1) - \alpha - \gamma, \quad (33)$$

$$0 = \frac{\partial L}{\partial x_1} = -(\ln x_1 + 1) - \alpha - \beta - -\gamma, \quad (34)$$

$$0 = \frac{\partial L}{\partial x_k} = -(\ln x_k + 1) - \alpha - k \cdot \beta \text{ for } k \le 2, \quad (35)$$

and defining $y = e^{-\beta}$ it follows that

$$\frac{x_1}{x_0} = y \quad \text{and} \quad x_k = x_2 \cdot y^{k-2}, \tag{36}$$

showing that the DPs again follow an exponential distribution. The difference to DPE1 is that the ratio $x_{k+1}/x_k = y$ is not observed for the ratio x_2/x_1. Constraints (13) and (31) imply

$$x_0 = \frac{p}{1+y} \quad \text{and} \quad x_2 = (1-p)(1-y). \tag{37}$$

Constraint (15) allows to derive the formula

$$b - 2(1-p) = p \cdot \frac{y}{1+y} + (1-p) \cdot \frac{y}{1-y}, \tag{38}$$

from which y can be determined from the initial conditions b (average number of bonds) and p (initially applied molar fraction of glucose and maltose).

For the special case $p = 0$ (in which no glucose or maltose is initially applied), Eq. (38) assumes the same form as for DPE1, with the exception that the left hand side reads $b - 2$. This is not surprising because under these conditions DPE2 will only mediate chemical transformations of glucans with at least two bonds (maltotriose). The dependency of the equilibrium parameter y on the average number of bonds, b and the initially applied amount of glucose and maltose, p is depicted in Figure 6.

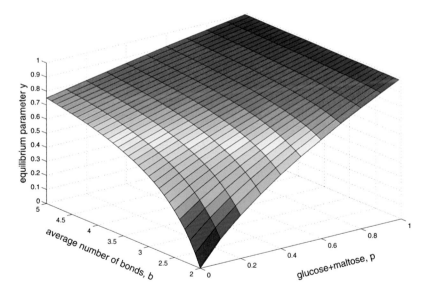

Figure 6. Predicted equilibrium parameter of DPE2 for varying average number of bonds (b) and for different amounts of glucose and maltose (p).

We have tested these predictions experimentally by incubating DPE2 with an initial mixture containing 40% maltose and 60% maltoheptaose [3]. These conditions correspond to the parameters $p = 0.4$ and $b = 4$. The experimentally observed equilibrium distribution and the theoretically predicted distribution for this initial condition is plotted in Figure 7.

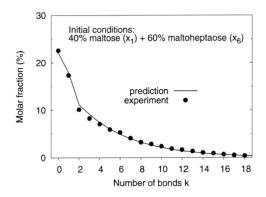

Figure 7. Predicted equilibrium distribution of the degree of polymerisation (solid line) and experimental validation (symbols) for DPE2. The reaction was initiated with a mixture of 40% (molar fraction) of maltose and 60% maltoheptaose, resulting in $p = 0.4$ and $b = 0.4 \times 1 + 0.6 \times 6 = 4$.

Phosphorylase

The enzyme α-glucan phosphorylase catalyses the transfer of a single glucose residue from the non-reducing end of a glucan onto orthophosphate to form glucose-1-phosphate. The general reaction is

$$P_i + G_n \leftrightarrow G1P + G_{n-1}, \tag{39}$$

Apparently, this enzyme also conserves the total number of molecules. However, since the bond enthalpies of the α-1,4 glucosidic linkages in polyglucans and the phosphoester bond in glucose-1-phosphate are different, the total energy of formation is not a conserved quantity. As a consequence, the equilibrium distribution will be determined by a combined effect of minimising the Gibbs energy of reaction and maximising the entropy. We denote by Δ_g the change in Gibbs energy when breaking one mole of α-1,4 glucosidic linkages and simultaneously forming one mole of phosphoester bonds. The molar fractions of orthophosphate (P_i) and of glucose-1-phosphate (G1P) are denoted by u and v, respectively. We assume that phosphorylase can be active on glucans with a minimal number of bonds, denoted m. As above, we denote with x_k the molar fraction of the glucan with k bonds. The total energy of formation of the reaction mixture (per mole) is thus

$$g^f = const. + v \cdot \Delta g \tag{40}$$

and the mixing entropy reads

$$S_{\text{mix}} = -R\left[u \ln u + v \ln v + \sum_{k \geq m} x_k \ln x_k\right]. \tag{41}$$

The equilibrium distribution is determined by identifying the minimum of the Gibbs free energy Eq. (11) under the constraints

$$\text{total number of molecules}: \quad u + v + \sum_{k \geq m} x_k = 1, \tag{42}$$

$$\text{conservation of bonds}: \quad v + \sum_{k \geq m} kx_k = b, \tag{43}$$

$$\text{conservation of phosphate groups}: \quad u + v = p. \tag{44}$$

These constraints are analogous to the three constraints (13), (15) and (31) which apply to DPE2. Indeed, they formally become identical if u is identified with x_0, v with x_1 and $m = 2$. The main difference is that here the Lagrange function

$$L(u, v, x_k; \alpha, \beta, \gamma) = v \cdot \Delta g + RT\left[u \ln u + v \ln v + \sum_k x_k \ln x_k\right] \tag{45}$$

$$+\alpha\left(u + v + \sum_k x_k - 1\right) + \beta\left(v + \sum_k kx_k - b\right) + \gamma(u + v - p)$$

contains the molar change in Gibbs energy Δ_g. Here, introducing

$$y = e^{-\frac{\beta}{RT}} \quad \text{and} \quad k_0 = e^{-\frac{\Delta g}{RT}} \tag{46}$$

and setting the partial derivatives to zero yields

$$\frac{v}{u} = y \cdot k_0 \quad \text{and} \quad \frac{x_{k+1}}{x_k} = y \quad \text{for} \quad k \geq m. \tag{47}$$

An analogous calculation to that performed for DPE2 yields

$$u = \frac{p}{1 + yk_0} \quad \text{and} \quad x_m = (1 - p)(1 - y) \tag{48}$$

and y is determined by solving the equation

$$b - m \cdot (1 - p) = p\frac{yk_0}{1 + yk_0} + (1 - p)\frac{y}{1 - y}. \tag{49}$$

The implicit formula (38) for DPE2 represents a special case of Eq. (49) when $k_0 = 1$, which corresponds to identical bond energies ($\Delta g = 0$). The analogous structure of the solutions is not surprising considering the parallels in the constraints that the respective enzymes observe. In both, the number of molecules as well as the number of bonds is conserved and both obey an additional, third, constraint. Whereas DPE2 conserves the sum of the glucose and maltose moieties, phosphorylase conserves the sum of the moieties of orthophosphate and glucose-1-phosphate.

The experimental validation of our predictions is illustrated in Figure 8. We have experimentally tested our predictions for two different initial conditions. We have initiated the reactions with a 1:4 mixture of glucans and glucose-1-phosphate, leading to $p = 0.8$. In one case, we chose maltoheptaose as the glucan, resulting in $b = 2.0$ (upper panel), whereas in the other case the glucan was maltotetraose, resulting in $b = 1.4$ (lower panel).

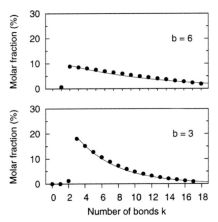

Figure 8. Predicted equilibrium distribution of the degree of polymerisation (solid line) and experimental validation (symbols) for phosphorylase. The reaction was initiated with a mixture of 20% (molar fraction) of glucan and 80% glucose-1-phosphate, resulting in $p = 0.8$. In the upper panel, maltoheptaose was chosen as glucan, leading to $b = 0.2 \times 6 + 0.8 \times 1 = 2.0$, for the lower panel the shorter glucan maltotetraose was used, resulting in $b = 0.2 \times 3 + 0.8 \times 1 = 1.4$.

Other glucosyltransferases

DPE2 and phosphorylase catalyse the reversible transfer of a terminal glucosyl residue onto an acceptor molecule. For DPE2, the acceptor is glucose, for phosphorylase, the acceptor is orthophosphate. Various other enzymes catalyse reactions according to this pattern. Most notably, starch synthases catalyse the elongation of glucans through the activated compound ADP-glucose. The general reaction scheme is

$$\text{ADPglc} + G_n \rightleftharpoons \text{ADP} + G_{n+1}. \tag{50}$$

Formally, ADP functions as the acceptor molecule even though here the equilibrium is far on the side of elongated chains and therefore transfer of glucosyl residues onto ADP are thermodynamically unfavourable and can be expected to occur with a slow rate. However, this reaction is structurally similar to reaction Eq. (39) and consequently the mathematical expression describing the equilibrium is also equivalent to Eq. (49). The only difference is the quantity of Δ_g, which for starch synthase describes the change in Gibbs free energy if one mole ADP-glucoses are split and one mole α-1,4 glucosidic linkages are formed. Similarly, glycogen synthase catalyses the elongation of glucans utilising UDP-glucose as donor molecule and the thermodynamic equilibrium distribution is expected to be highly similar to that of starch synthase.

The enzyme amylosucrase parallels DPE2 in that it utilises fructose (Fru) as an acceptor molecule, forming sucrose (Suc) according to the scheme

$$G_n + \text{Fru} \rightleftharpoons G_{n-1} + \text{Suc}. \tag{51}$$

Again, the reaction is structurally similar to reaction Eq. (39) and the mathematical formalism to determine the equilibrium is identical.

EXPERIMENTAL DETERMINATION OF BOND ENERGIES

The analytic calculations of equilibrium distributions for different glucosyltransferases have revealed interesting parallels. All reaction systems can be written in the form

$$u + x_k \rightleftharpoons v + x_{k-1} \quad \text{with } k \geq m. \tag{52}$$

In the forward direction (left to right), a glucan (x_k) is shortened by one glucosyl residue, which is transferred to an acceptor molecule u. In the opposite direction, the bound form v acts as a glucosyl donor, transferring the glucosyl residue back to a glucan (x_{k-1}), thereby elongating it. The minimal number of bonds in a glucan in order to act as a substrate is denoted by m. For reaction systems of this type, the sum of u and v is a conserved quantity, $u + v = p$.

For DPE2, u and v are identified with glucose (x_0) and maltose (x_1), respectively, and $m = 2$, reflecting the separation of the glucose/maltose pool from glucans with higher DP. In the case of amylosucrase, u is to be identified with fructose and v with sucrose. For phosphorylase, u and v are identified with orthophosphate (P_i) and glucose-1-phosphate (G1P), respectively. For starch synthase u stands for ADP and v for ADP-glucose. While the mechanisms of all three enzymes are quite different, their catalytic activity can formally be written in the same way (see Eq. 52) and the only difference is in their Gibbs free energy of reaction.

It is therefore worthwhile studying Eq. (49) to investigate the effect of Δ_g on the equilibrium distribution to develop a method by which Δ_g can be determined from experimentally observed equilibria. First, it is important to note that the left-hand side of Eq. (49) denotes the average number of *active* bonds in the reactant mixture, because b is the total average number of bonds and $m(1 - p)$ is the average number of *inactive* bonds ($1 - p$ is the molar fraction of all glucans with m or more bonds, and in every such glucan these m bonds are never subject to enzymatic hydrolysis). We therefore introduce the quantity b' reflecting the number of active bonds by letting

$$b' = b - m \cdot (1 - p). \tag{53}$$

We now consider the extreme case of an irreversible reaction (52) running in the direction right to left. The equilibrium is completely on the left side of Eq. (52), thus no v is present ($v = 0$ and $u = p$). Under these conditions, $\Delta g \to \infty$, implying $k_0 \to 0$ and Eq. (49) becomes

$$\frac{b'}{1-p} = \frac{y}{1-y}. \tag{54}$$

This expression is exactly analogous to Eq. (23), determining the equilibrium parameter y for DPE1 for an average number of bonds $b'/(1 - p)$, which corresponds to the average number of active bonds in the glucans.

The other extreme situation of irreversibility in the direction left to right in Eq. (52) is described by $\Delta g \to -\infty$, implying $k_0 \to \infty$. Under these circumstances, glucan chains will be maximally shortened and the glucosyl residues will be transferred to the donor u to form v. Here, we need to distinguish two cases: If there exist more active bonds in the system than acceptor molecules, the equilibrium will be characterized by an absence of acceptor molecules ($u = 0$ and $v = p$). If, however, there are fewer transferrable glucosyl residues than acceptor molecules, some unbound acceptor molecules remain in the system ($u > 0$). The first scenario is observed when $b' > p$, because then there exist more active bonds than acceptor molecules. In this case, the equation (49) determining the equilibrium parameter y becomes

$$\frac{b'-p}{1-p} = \frac{y}{1-y}. \tag{55}$$

Again, this expression is exactly analogous to Eq. (23) for DPE1, now with an average number of bonds $(b' - p)/(1 - p)$. The difference to Eq. (54) results from the fact that here p bonds are present in the molecules v.

The second scenario represents a situation with an insufficient number of bonds ($b' < p$). Eq. (55) cannot be used to determine the equilibrium distribution in this case because no positive solution for y exists. The only non-negative solution of Eq. (49) for $k_0 \to \infty$ is given by $y = 0$, where

$$\lim_{k_0 \to \infty} y k_0 = \kappa \tag{56}$$

remains finite. Then,

$$b' = p \cdot \frac{\kappa}{1+\kappa} = v, \tag{57}$$

and the equilibrium distribution is monodisperse with

$$x_m = 1 - p \text{ and } x_k = 0 \text{ for } k \geq m. \tag{58}$$

These theoretical deliberations allow the following conclusion: In a situation with a small average number of active bonds and a large excess of glucosyl acceptors u, the equilibrium parameter y will strongly depend on Δ_g, representing the change in Gibbs energy when converting α-1,4 glucosidic linkages in bonds linking glucosyl residues to form v. Let us, for example, assume a 1:50 mixture of glucans and glucosyl acceptors, where the glucans contain on average three active bonds. This means, that the total average number of bonds in a glucan is $m + 3$, where m is the number of inactive bonds. Then, $p = 50/51$ and $b' = 3/51 < p$. Together with Eq. (54) the limiting cases read

$$\lim_{\Delta g \to \infty} y = \frac{3}{4} \quad \text{and} \quad \lim_{\Delta g \to -\infty} y = 0. \tag{59}$$

These extreme values allow to experimentally determine Δ_g from the equilibrium distribution of the glucans. The equilibrium parameter can be directly read out from the equilibrium distribution since in a semi-logarithmic plot of the molar fractions of the glucans the slope amounts to $-\ln(y)$.

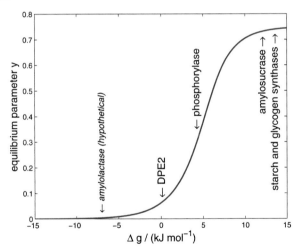

Figure 9. The equilibrium parameter y in dependence on the difference in Gibbs free energies. The chosen parameters are $b' = 3/51$ and $p = 50/51$. The values for the glucosyltransferases DPE2 ($\Delta g = 0$), phosphorylase ($\Delta g = 4.1$ kJ/mol), amylosucrase ($\Delta g = 12$ kJ/mol) and starch/glycogen synthase ($\Delta g = 13.5$ kJ/mol) are indicated. Further, a hypothetical enzyme, *amylolactase* with a negative value ($\Delta g = -7.1$ kJ/mol) is included for illustration.

Figure 9 depicts the dependence of the equilibrium parameter y on the change in Gibbs free energy Δ_g for these parameter values. The predicted equilibrium distributions for the glycosyltransferases discussed in this paper are denoted in the plot. For DPE2, $\Delta g = 0$. For phosphorylase, we previously predicted with this method a value of $k_0 = 0.19$ [3], corresponding to $\Delta g = +4.1$ kJ/mol. For amylosucrase, we estimated the equilibrium constant using the eQuilibrator web interface [10] as $\Delta g = +12$ kJ/mol. For starch synthase, this method estimated a value of $\Delta g = +13.5$ kJ/mol.

Additionally, a hypothetical enzyme with negative Δ_g is indicated. In analogy to amylosucrase, which transfers the terminal glucosyl residue from a glucan to fructose forming sucrose, a hypothetical amylolactase would catalyse the reaction

$$G_n + \text{galactose} \; G_{n-1} + \text{lactose} \tag{60}$$

with an estimated value $\Delta g = -7.1$ kJ/mol.

CONCLUSIONS

We have presented a concise thermodynamic characterization of enzymatic reactions on carbohydrate polymers. The description of the action of these carbohydrate-active enzymes by concepts from statistical thermodynamics allows to precisely predict and characterise the corresponding equilibrium distributions. We introduced the equilibrium parameter β, or equivalently $y = \exp(-\beta/RT)$, which fully charaterizes the distributions of degrees of polymerisation in equilibrium and thus represents a suitable generalisation of the classical equilibrium constant K_{eq} for carbohydrate-active enzymes.

The theory further provides a method how differences in bond enthalpies can be experimentally determined through the observation of equilibrium distributions. For this, an excess of acceptor molecules and rather short chain lengths should be chosen as initial conditions, which leads to a strong dependence of the equilibrium parameter on the bond enthalpy difference Δ_g.

Due to the generality of our theory, our new concepts are not limited to the specific enzymes discussed in the text or even to carbohydrate-active enzymes. We expect that our thermodynamic approach is widely applicable to many polymer-active enzymes in general.

REFERENCES

[1] Ball, S.G. and Morell, M.K. (2003) From bacterial glycogen to starch: understanding the biogenesis of the plant starch granule. *Annu. Rev. Plant Biol.* **54**:207 – 233. doi: http://dx.doi.org/10.1146/annurev.arplant.54.031902.134927.

[2] Goldberg, R.N., Bell, D., Tewari, Y.B. and McLaughlin, M.A. (1991) Thermodynamics of hydrolysis of oligosaccharides. *Biophys. Chem.* **40**(1):69 – 76. doi: http://dx.doi.org/10.1016/0301-4622(91)85030-T.

[3] Kartal, Ö., Mahlow, S., Skupin, A. and Ebenhöh, O. (2011) Carbohydrateactive enzymes exemplify entropic principles in metabolism. *Mol. Syst. Biol.* **7**:542. doi: http://dx.doi.org/10.1038/msb.2011.76.

[4] Alberty, R.A. (2003) Thermodynamics of Biochemical Reactions. John Wiley & Sons, Hoboken. doi: http://dx.doi.org/10.1002/0471332607.

[5] Schuster, S. and Höfer, T. (1991) Determining all extreme semi-positive conservation relations in chemical reaction systems: a test criterion for conservativity. *J. Chem. Soc., Faraday Trans.* **87**:2561 – 2566. doi: http://dx.doi.org/10.1039/ft9918702561.

[6] Jones, G. and Whelan, W.J. (1969) The action pattern of d-enzyme, a transmaltodextrinylase from potato. *Carbohydrate Research* **9**(4):483 – 490. doi: http://dx.doi.org/10.1016/S0008-6215(00)80033-6.

[7] Lin, T.P., Spilatro, S.R. and Preiss, J. (1988) Subcellular localization and characterization of amylases in arabidopsis leaf. *Plant Physiol.* **86**(1):251 – 259. doi: http://dx.doi.org/10.1104/pp.86.1.251.

[8] Colleoni, C., Dauvillée, D., Mouille, G., Morell, M., Samuel, M., Slomiany, M.-C., Liénard, L., Wattebled, F., d'Hulst, C. and Ball, S. (1999) Biochemical characterization of the *Chlamydomonas reinhardtii* alpha-1,4 glucanotransferase supports a direct function in amylopectin biosynthesis. *Plant Physiol.* **120**(4):1005 – 1014. doi: http://dx.doi.org/10.1104/pp.120.4.1005.

[9] Steichen, J.M., Petty, R.V. and Sharkey, T.D. (2008) Domain characterization of a 4-alpha-glucanotransferase essential for maltose metabolism in photosynthetic leaves. *J. Biol. Chem.* **283**(30):20797 – 20804. doi: http://dx.doi.org/10.1074/jbc.M803051200.

[10] Flamholz, A., Noor, E., Bar-Even, E. and Milo, R. (2012) eQuilibrator – the biochemical thermodynamics calculator. *Nucleic Acids Res.* **40**(Database issue): D 770 – D 775. doi: http://dx.doi.org/10.1093/nar/gkr874.

Beilstein-Institut

METAL BINDING SITES IN PROTEINS

VLADIMIR SOBOLEV*, RONEN LEVY, MARIANA BABOR AND MARVIN EDELMAN[#]

Department of Plant Sciences, Weizmann Institute of Science, Rehovot, Israel

E-MAIL: *vladimir.sobolev@weizmann.ac.il and
[#]marvin.edelman@weizmann.ac.il

Received: 22nd February 2012/Published: 15th February 2013

ABSTRACT

Metal ions play a critical role in living systems. About one third of
proteins need to bind metal for their stability and/or function. In this
review, current sequence based and structure based methods for metal
binding site prediction will be presented, with emphasis on the CHED
and SeqCHED methods of prediction from apo-protein structures and
protein sequences having homologs (even remote) in the structural
protein databank (PDB). Metal binding site prediction will be
considered as a step in function assignment for new proteins. Finally,
a disproportional association of first and second shell metal binding
residues in human proteins with disease-related SNPs will be shown.

INTRODUCTION

Biological cells must adapt strict regulatory mechanisms in order to maintain metal homeostasis within the cytoplasm [1]. While metal ions can be utilized in various manners in a
biological system, the position of a metal ion in space, its variation in time, and the exact
chemical partner with which it interacts (often a protein) have been selected by the demands
of evolution [2].

Metal ions are required for a great variety of purposes in proteins and are present in more
than one third of protein structures investigated [3, 4]. Metals increase the structural stability
of the protein in the proper conformation required for biological function. A metal ion can
serve as a cross-linking agent, since metal ions usually bind through several interactions with

amino acid side chains [2]. In addition, metals can be directly involved in the chemical reactions catalysed by an enzyme. They can serve as redox centres for catalysis (e. g., haem-iron centres) or as electrophilic reactants in catalysis [5]. Metals can help to bring reactive groups into the correct orientation for reaction.

Furthermore, metals can play a regulatory role in proteins. This includes a role in signal transduction, in controlling the architecture of protein complexes, and in redox-active metal sites, where the binding and release of the metal is under redox control [6]. Metals have several valence states, which depending on their ligands can lie close in energy. As a consequence, the metal can be switched from state to state upon binding to a protein, resulting in considerable protein changes [2]. Because of the above, it is important to be able to predict metal binding site based on sequence and/or structural information.

SEQUENCE-BASED METAL BINDING SITE PREDICTIONS

One of the approaches [7], taken to harvest sequence information, systematically determines all possible metal-binding signatures present in the Protein Data Bank. These signatures, termed MBP (Metal Binding Patterns), include the binding residues and their spacing along the sequence. The method was applied to copper proteins, and a library of metal binding patterns was built. Each MBP is used together with the primary sequence of the corresponding metalloprotein to browse any ensemble of sequences of interest. The level of confidence of this method is variable, ranging between 50% to over 90%, depending on the lengths of the local alignments identified around each binding residue. As this work was applied only to copper, it is not clear to what extent it is applicable to other metals. Moreover, a limitation of this work is that it requires identification of conserved spacing patterns between binding residues and these spacings are not conserved in all cases. Hence, it is not possible to search for a binding residue that is far away in sequence from other binding residues, since the exact spacing can vary greatly among sequences. In another study [8], multiple sequence alignment, entropy (residue conservation) and relative weight of gapless matches were obtained, and the correlation between nearby residues was modelled by support vector machine semi-pattern predictors.

Another algorithm [5] takes subsequences of proteins as input, under the assumption that metal binding residues are influenced by the surrounding environment in nature. The amino acid at the centre of the fragment is the target amino acid, whereas the others are the "neighbours". The fragment sequence is encoded to a feature vector, which contains information on the occurrence probability of the amino acid, the propensities of the secondary structure, and the metal-binding propensity of the amino acid. The feature vector is fed into a neural-network learning machine. The learning machine decides whether the target amino acid binds metal or not. This process is repeated by shifting each time one position along the protein sequence, resulting in a new fragment. With this algorithm, binding residues are identified with higher than 90% sensitivity. However, the limitation of this approach is that it predicts metal binding residues rather than metal binding sites. Therefore, it analyses the probability of each putative binding residue individually, instead of taking into consideration the combined context of all residues belonging to one unified site. In some proteins one protein chain can include more than one binding site (for example, 14% of zinc binding sites fall in this category). Thus, the binding residues of different sites can be erroneously intertwined [6].

A third algorithm [9] scans the sequence around the four main residue types involved in metal binding (Cys, His, Asp, Glu; [10 – 12]) using a window of up to 25 residues, physicochemical features (including conservativity) and correlated mutation analysis derived from multiple sequence alignment.

STRUCTURE-BASED METAL BINDING SITE PREDICTIONS

One of the first algorithms [13] is based on the finding that many metal sites in proteins share a common feature: they are cantered in a shell of hydrophilic ligands, surrounded by a shell of carbon-containing groups. Therefore, it is possible to measure the contrast between groups located at the centre of the sphere (more hydrophilic), and groups located at the outer shell (more hydrophobic) within a radius threshold distance. The contrast function is generally maximal when cantered at or near a metal binding site. However, this algorithm also identified regions of high contrast that were not associated with metal binding, such as charged surface residues and buried, positively-charged residues [14].

A second algorithm [15] is designed specifically for Ca^{2+} binding site prediction, since it is based on the finding that the coordination shell of Ca^{2+} ions in proteins contains almost exclusively oxygen atoms supported by an outer shell of carbon atoms. The bond strength contribution of each ligating oxygen in the inner shell can be evaluated, and the sum of such contributions closely approximates the valence of the bound cation. Assuming local neutralization of charges, the bond strength, or bond order, contributed by each oxygen ligand to the ligated cation is the charge of the cation divided by the number of ligands, or the coordination number. When ligands are asymmetrically disposed around the ligating cation, different bonds are expected to have different strengths. Here, the bond-length correlation to bond order, which is also seen in covalent bonding, can be used to estimate the strength of different bonds in structures. When a protein is embedded in a very fine grid of points and an algorithm is used to calculate the valence of each point (representing a potential binding site), a typical distribution of valence values is obtained. However, only a very small fraction of the points have a significantly large valence value. These points share a tendency to cluster around known Ca^{2+} ions, enabling prediction of such sites.

Sodhi *et al.* [16] calculated the likelihood of a given residue to be a metal ligand by considering multiple sequence alignment of homologous proteins as well as approximate structural information. This method, called MetSite, performed satisfactorily for SCOP database superfamilies [17] where large sets of evolutionary related proteins are available. The algorithm was developed considering 190, 18, 11 and 49 superfamilies for Zn, Fe, Cu and Mn, respectively, while valuation of performance was applied to five, four, one and one cases, respectively. As with Lin *et al.* (2005), MetSite suffers from difficulties to identify the location of a metal binding site by inspecting the distribution of predicted individual residues within the protein structure.

The Fold-X algorithm [18] specializes in predicting the spatial position of a metal in the protein. It uses a library, extracted from the PDB, containing the most common metal spatial positions relative to the corresponding ligating atoms. In the first step, this library is used to search for possible metal positions within the protein structure. Then, an optimization step is performed to find the best position for the predicted metal using the Fold-X force field [19]. The resulting position is used to estimate the energy of binding. At the end, a hydration step to add water ligands is also included. This algorithm is geared to, and performed well in identifying the position of metals in holo forms.

FEATURE, a machine learning method based on a Bayesian classifier was used to identify zinc and calcium binding sites in proteins [20, 21]. This method uses many averaged biochemical and biophysical features in six concentric spherical shells around a suspected site. Shell features include number of atoms, Van der Waals volume, hydrophobicity, solvent accessibility, the presence of different oxygens, nitrogens, carbons and sulphur atoms, amino acid residues, and charges. Similar to Fold-X, FEATURE predicts the position of metal ions within the predicted binding site.

CHED METAL BINDING SITE PREDICTION

As mentioned previously, it is well established that four residues: Cys (C), His (H), Glu (E) and Asp (D) (referred to as "CHED" by Babor *et al.* [22]), are the most common amino acids forming soft metal binding sites [10 – 12]. The CHED prediction algorithm [22] is composed of two steps. Step 1 is based on a statistical comparison of holo and apo structure pairs, which showed that at most one ligand side chain reorients upon metal binding [23, 24]. In this step (Fig. 1), the algorithm searches for a 3D constellation of three amino acid residues, whose metal-ligating atoms satisfy distance criteria and where at most one side chain has rotated among the three residues. A binding site is defined as a single triad, or multiple triads that share at least one residue between two or more of them. The second step involves filtration and eliminates false positives. A "mild" filter was created based on the observation that sites composed of a large number of triads tend to be true. Therefore, in cases where a site is found to contain at least five triads, all other putative sites with three or fewer triads are discarded. This filtration deletes about 10% of metal binding sites in apo proteins, yielding a sensitivity (percentage of correctly predicted experimentally known metal binding sites) of 90%. However, among binding sites predicted, 38% proved to be false positives, yielding a selectivity (percentage of correct binding sites among all predicted) of 62%.

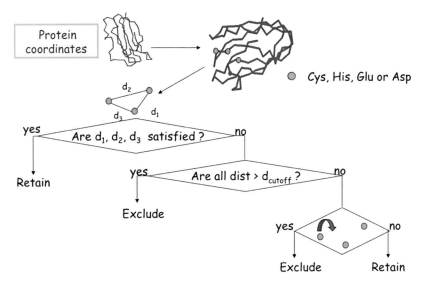

Figure 1. Schematic presentation of step 1 in the CHED algorithm. All possible sets of three amino acid residues (triads) from the four CHED residues, whose collective distances of Cβ atoms are less than 13.0 Å were retrieved. A triad was retained if distances d_1, d_2 and d_3 among ligand atoms from separate CHED residues satisfied individual cutoff criteria. These cutoff values were chosen by analyzing a large (over 1000 sites), redundant set of holo forms and refined using available apo structures. In addition, if one or two out of the three inter-ligand distances were not initially satisfied, alternative side chain conformations of the relevant residues were built, one at the time, using a backbone-independent rotamer library [25]. If no clashes were eventually observed, and d_1, d_2 and d_3 now satisfied the cutoff distances, then the built up triad was retained.

To increase selectivity, a "stringent" filter was created using a decision tree with the following features: number of times a residue of a potential binding site is selected (since a specific residue can belong to more than one initial 'binding site' before joining them together); proportions between C, H, E, D amino acids of a potential binding site; number of sites predicted for the protein; residue sequence entropy; hydrogen bond surface areas between the potential binding residues and any of its neighbouring amino acids. Furthermore, a support vector machine classifier was added, which included the above parameters plus the number of triads per predicted site and relative solvent accessible surface. Triads excluded by both the decision tree and support vector machine classifier were removed. Stringent filtration reduced sensitivity to about 70%. Importantly, it upped selectivity to 90%.

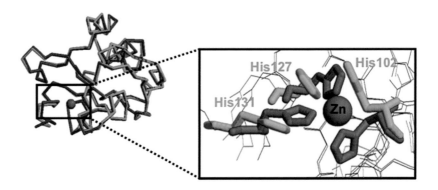

Figure 2. Superimposition of the holo (magenta) and apo (green) colicin E9 DNase domains. The ligand residues for zinc in the apo form were correctly predicted, even though the maximum Cα displacement was equal to 2.45 Å and rearrangement of His102 upon metal ion binding was observed. The coordinates for chains B were taken from PDB entries 2gze (holo form) and 1env (apo form) PDB entries. Binding site residues were found using LPC software [26].

The search procedure for sites has sufficient flexibility built in to often allow for some backbone shifts as well as side chain reorientations. An example is the Colicin E9 DNase domain (Fig. 2). Here the entire binding site was identified successfully in the apo form.

SEQCHED METAL BINDING SITE PREDICTION

We developed SeqCHED [27] for prediction of metal binding sites starting with translated DNA sequence data. The method integrates three tools: PsiBlast, SCCOMP and CHED (Fig. 3). PsiBlast [28] identifies statistically significant alignments using a position-specific score matrix that is derived iteratively. The tool was used in a specific manner: target sequences were first subjected to two iterations of PsiBlast against the NCBI non-redundant protein sequence database (NR, http://www.ncbi.nlm.nih.gov/blast/blast_databases.shtml) with a third iteration against PDB to identify structural templates. SCCOMP [29] is a method for side chain modelling. It uses a scoring function including terms for complementarity, excluded volume, internal energy based on roamer probability and solvent accessible surface. The CHED procedure was described above. Table 1 summarizes statistics for SeqCHED prediction. Again, importantly, upon stringent filtration selectivity is higher than 85%.

Table 1. Predictability of transition metal binding sites in modeled structures

PDB template	No. of modeled sites	Sequence identity	Mild filter		Stringent filter	
			% Sensitivity	% Selectivity	% Sensitivity	% Selectivity
Metal containing	223	Target (native)	98	63	93	92
	223	Target (self model)	95	57	84	92
	202	30–99%	95	58	84	93
	98	18–30%	86	53	85	82
Non-metal containing	143	Target (native)	91	61	76	89
	143	Target (self model)	90	54	67	86
	99	30–99%	79	52	49	89
	162	18–30%	65	42	33	90

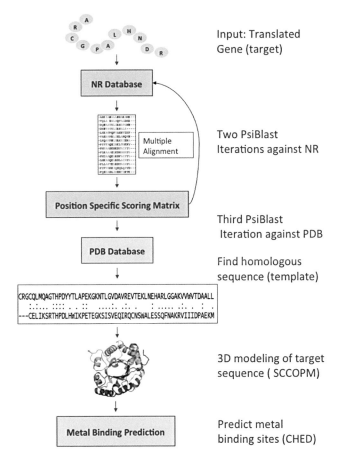

Input: Translated Gene (target)

Two PsiBlast Iterations against NR

Third PsiBlast Iteration against PDB

Find homologous sequence (template)

3D modeling of target sequence (SCCOPM)

Predict metal binding sites (CHED)

Figure 3. Scheme of the SeqCHED procedure. The procedure includes two Psi-Blast iterations against a non-redundant sequence database, a third iteration against the PDB database (to find a structural template), 3D modeling of the target sequence (using SCCOMP for side chain placement), followed by metal binding site prediction using the CHED algorithm.

LINKAGE BETWEEN DISEASE-ASSOCIATED SNPs AND METAL BINDING SITES

We recently found that mutations associated with diseases (protein variants) are associated with metal binding sites significantly more often than expected [30]. Among the sequences having disease-related single nucleotide polymorphisms (dSNPs), 40% involve mutation of a CHED residue, while for sequences not associated with disease (ndSNPs) the level is 30%. This difference is highly significant and suggests a bias for association of dSNPs with metal binding sites. An analysis of the relation between dSNPs and metal binding sites is presented in Fig. 4. The results demonstrate a clear bias of dSNPs in the immediate vicinity of metal binding sites.

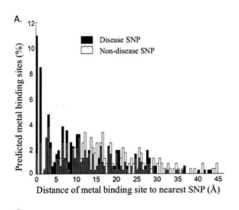

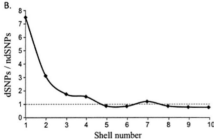

Figure 4. Proximity of dSNPs to metal binding sites (Figure from Levy *et al.* [30]). All proteins containing predicted metal binding sites derived from the Human Polymorphisms and Disease Mutations index were analyzed. The data sets are composed of 237 sequences containing 320 predicted sites with one or more dSNPs, and 184 sequences containing 243 predicted sites with one or more ndSNPs. **A.** The histogram shows the percent distribution of distances between predicted site residues and the nearest dSNP (black bars) or ndSNP (white bars). The overlap between the two distributions is colored gray. The first bar represents the predicted binding site residues (first shell); the bar between 1.0 and 1.5 Å, covalently bound second shell residues; the bars between 2 Å and 4.5 Å, non-covalently bound second shell residues. Bars at greater distances represent residues in successively remote shells. **B.** The normalized ratio between the number of all dSNPs and ndSNPs was obtained for 10 successive shells. A clear differential between the number of dSNPs and ndSNPs can be seen for the first and second shells, and to a lesser extent, for the third and forth shells. The curve reaches a plateau (dashed line) at the fifth shell.

REFERENCES

[1]   Coombs, J.M. and Barkay, T. (2005) New findings on evolution of metal homeostasis genes: Evidence from comparative genome analysis of bacteria and archaea. *Appl. Environ. Microbiol.* **71**:7083 – 7091.
doi: 10.1128/AEM.71.11.7083-7091.2005.

[2]   Williams, R.J.P (1985) The symbiosis of metal and protein functions. *Eur. J. Biochem.* **150**:213 – 248.
doi: 10.1111/j.1432-1033.1985.tb09013.x.

[3]   Ibers, J.A. and Holm, R.H. (1980) Modeling coordination sites in metallobiomolecules. *Science* **209**:223 – 235.
doi: 10.1126/science.7384796.

[4]   Tainer, J.A., Roberts, V.A., and Getzoff, E.D. (1992) Protein metal-binding sites. *Curr. Opin. Biotechnol.* **3**:378 – 387.
doi: 10.1016/0958-1669(92)90166-G.

[5]   Lin, C.T., Lin, K.L., Yang, C.H., Chung, I.F., Huang, C.D., Yang, Y.S. (2005) Protein metal binding residue prediction based on neural networks. *Int. J. Neur. Syst.* **15**:71 – 84.
doi: 10.1142/S0129065705000116.

[6]   Maret, W. (2005) Fluctuations of cellular, available zinc modulate insulin signaling via inhibition of protein tyrosine phosphatases. *J. Trace Elem. Med. Biol.* **19**:7 – 12.
doi: 10.1016/j.jtemb.2005.02.003.

[7]   Andreini, C., Bertini, I., Rosato, A. (2004) A hint to search for metalloproteins in gene banks. *Bioinformatics* **20**:1373 – 1380.
doi: 10.1093/bioinformatics/bth095.

[8]   Passerini, A., Andreini, C., Menchetti, S., Rosato, A., Frasconi, P. (2007) Predicting zinc binding at the proteome level. *BMC Bioinf.* **8**, Article Number 39.

[9]   Shu, N., Zhou T., Hovmoller S. (2008) Prediction of zinc-binding sites in proteins from sequence. *Bioinformatics* **24**:775 – 782.
doi: 10.1093/bioinformatics/btm618.

[10]   Alberts, I.L., Nadassy, K., Wodak, S. J. (1998) Analysis of zinc binding sites in protein crystal structures. *Prot. Sci.* **7**:1700 – 1716.
doi: 10.1002/pro.5560070805.

[11]   Auld, D. S. (2001) Zinc coordination sphere in biochemical zinc sites. *Biometals* **14**:271 – 313.
doi: 10.1023/A:1012976615056.

[12] Dudev, T., Lin, Y.L., Dudev, M., Lim, C. (2003) First-second shell interactions in metal binding sites in proteins: a PDB survey and DFT/CDM calculations. *J. Amer. Chem. Soc.* **125:**3168 – 3180.
doi: 10.1021/ja0209722.

[13] Yamashita, M.M., Wesson, L., Eisenman, G., Eisenberg, D. (1990) Where metal-ions bind in proteins. *Proc. Natl. Acad. of Sci. U.S.A.* **87:**5648 – 5652.
doi: 10.1073/pnas.87.15.5648.

[14] Gregory, D.S., Martin, A.C., Cheetham, J.C., Rees, A.R. (1993. The prediction and characterization of metal binding sites in proteins. *Prot. Eng.* **6:**29 – 35.
doi: 10.1093/protein/6.1.29.

[15] Nayal, M. and DiCera, E. (1994) Predicting Ca^{2+}-binding sites in proteins. *Proc. Natl. Acad. of Sci. U.S.A.* **91:**817 – 821.
doi: 10.1073/pnas.91.2.817.

[16] Sodhi, J.S., Bryson, K., McGuffin, L.J., Ward, J.J., Wernisch, L., Jones, D.T. (2004). Predicting metal-binding site residues in low-resolution structural models. *J. Mol. Biol.* **342:**307 – 320.
doi: 10.1016/j.jmb.2004.07.019.

[17] Murzin, A.G., Brenner, S.E., Hubbard, T., Chothia, C. (1995) SCOP: a structural classification of proteins database for the investigation of sequences and structures. *J. Mol. Biol.* **247:**536 – 540.
doi: 10.1016/S0022-2836(05)80134-2.

[18] Schymkowitz, J.W., Rousseau, F., Martins, I.C., Ferkinghoff-Borg, J., Stricher, F., Serrano, L. (2005) Prediction of water and metal binding sites and their affinities by using the Fold-X force field. *Proc. Natl. Acad. Sci. U.S.A.* **102:**10147 – 10152.
doi: 10.1073/pnas.0501980102.

[19] Guerois, R., Nielsen, J.E., Serrano, L. (2002) Predicting changes in the stability of proteins and protein complexes: a study of more than 1000 mutations. *J. Mol. Biol.* **320:**369 – 387.
doi: 10.1016/S0022-2836(02)00442-4.

[20] Ebert, J.C. and Altman, R.B. (2008) Robust recognition of zinc binding sites in proteins. *Prot. Sci.* **17:**54 – 65.
doi: 10.1110/ps.073138508.

[21] Liu, T. and Altman R.B. (2009) Prediction of calcium-binding sites by combining loop-modeling with machine learning. *BMC Struc. Biol.* **9:**72.
doi: http://doix.org/10.1186/1472-9-72.

[22] Babor, M., Gerzon, S., Raveh, B., Sobolev, V, Edelman, M. (2008) Prediction of transition metal-binding sites from apo protein structures. *Proteins* **70**:208–217. doi: 10.1002/prot.21587.

[23] Babor, M., Greenblatt, H.M., Edelman, M., Sobolev, V. (2005) Flexibility of metal binding sites in proteins on a database scale. *Proteins* **59**:221–230. doi: 10.1002/prot.20431.

[24] Edelman, M., Babor, M., Levy, R., Sobolev, V. (2008) Metalloproteins: Structure, conservation and prediction of metal binding sites. In *Structural proteomics and its impact on the life sciences*. eds. Sussman, J.L. and Silman, I., pp. 181–205.

[25] Dunbrack, R.L., Jr., and Cohen, F.E. (1997) Bayesian statistical analysis of protein side-chain rotamer preferences. *Prot. Sci.* **6**:1661–1681. doi: 10.1002/pro.5560060807.

[26] Sobolev V., Sorokine A., Prilusky J., Abola E.E., Edelman M. (1999) Automated analysis of interatomic contacts in proteins. *Bioinformatics* **15**:327–332. doi: 10.1093/bioinformatics/15.4.327.

[27] Levy, R., Edelman, M., Sobolev, V. (2009) Prediction of 3D metal binding sites from translated gene sequences based on remote-homology templates. *Proteins* **76**:365–374. doi: 10.1002/prot.22352.

[28] Altschul, S.F., Madden, T.L., Schaffer, A.A., Zhang, J.H., Zhang, Z., Miller, W., Lipman, D.J. (1997) Gapped BLAST and PSI-BLAST: a new generation of protein database search programs. *Nucl. Acids Res.* **25**:3389–3402. doi: 10.1093/nar/25.17.3389.

[29] Eyal, E., Najmanovich, R., Mcconkey, B.J., Edelman, M., Sobolev, V. (2004) Importance of solvent accessibility and contact surfaces in modeling side-chain conformations in proteins. *J. Comput. Chem.* **25**:712–724. doi: 10.1002/jcc.10420.

[30] Levy, R., Sobolev, V., Edelman, M. (2011) First- and second-shell metal binding residues in human proteins are disproportionately associated with disease-related SNPs. *Hum. Mutation* **32**:1309–1318. doi: 10.1002/humu.21573.

Biographies

Karen Allen

received her B.S. degree in Biology, cum laude from Tufts University and her Ph.D. in Biochemistry from Brandeis University, where she was a Dretzin scholar. Her graduate studies in the laboratory of the mechanistic enzymologist, Dr. Robert H. Abeles, focused on the design, synthesis, and inhibition kinetics of transition-state analogues. Following her desire to see enzymes in action she pursued X-ray crystallography during postdoctoral studies as an American Cancer Society Fellow in the laboratory of Drs. Gregory A. Petsko and Dagmar Ringe. Since 1993 she has lead her own research team at Boston University, first in the Department of Physiology and Biophysics at the School of Medicine, and since 2008 in the Department of Chemistry where she is now a Professor. She is also on the faculty of both the Bioinformatics and Cell and Molecular Biology programs at Boston University. Dr. Allen's research has focused on the elucidation of enzyme mechanisms and the understanding of how Nature has evolved new chemistries from existing protein scaffolds. Within this context, her laboratory has plumbed the basis of enzyme-mediated phosphoryl transfer. Her laboratory is currently incorporating the lessons learned from superfamily analysis to the assignment of function as part of the Enzyme Function Initiative.

In addition, Dr. Allen has sought to provide new tools for the exploration of protein structure and function by the invention and implementation of lanthanide binding tags. Dr. Allen's students and postdoctoral researchers have gone on to research positions in structural genomics institutes such as RIKEN, Japan and drug discovery companies including AstraZeneca and Novartis as well as in the academic arena as independent investigators. She is an Associate Editor of the American Chemical Society journal Biochemistry, and on the Editorial Advisory Board of Chemical Reviews and has served on both NIH and NSF study sections as a regular panel member. Dr. Allen has been an invited lecturer and seminar speaker on over sixty occasions, and has chaired a number of national and international meetings.

Antonio Baici

studied chemistry and obtained his doctorate at the University of Trieste (Italy) in 1970. After completion of a post-doctoral period at the Swiss Federal Institute of Technology in Zurich (ETH), he continued at the same institution working on thermodynamics and kinetics of dehydrogenases and other enzymes, as well as on conformational properties of oligopeptides using NMR methods. Between 1978 and 2001 he was group leader at the

Department of Rheumatology of the University Hospital in Zurich and focused on the degradation of connective tissue by peptidases in rheumatic diseases and tumor invasion. He proposed a mechanism for the pathogenesis of osteoarthritis based on the transition of chondrocytes to a dedifferentiated phenotype, and demonstrated a pivotal role of cathepsin B in maintaining the chronicity of the disease. He was appointed Associate Professor in 1989, Professor of Biochemistry in 1996 and in 2001 he moved to the Department of Biochemistry of the University of Zurich.

Since 1980 emphasis was put on understanding how proteolytic enzymes interact with components of the extracellular matrix at the cell-matrix interface in physiological processes, and why naturally-occurring inhibitors are not able to counteract abnormal matrix degradation under pathological conditions. While investigating the action of inhibitors aimed at counteracting unwanted proteolysis, several enzyme kinetic methods were developed for the diagnosis and quantification of hyperbolic mixed-type inhibitors, allosteric effectors and the simultaneous interaction of enzymes with two modifiers. After retiring in May 2011, he can at last dedicate plenty of time to his favored activities, enzyme kinetics and violin making.

Oliver Ebenhöh

08/2009 –	Reader (associate professor) in Systems Biology, Coordinator of the Theoretical Systems Biology Research Programme, Institute for Complex Systems and Mathematical Biology, University of Aberdeen
03/2007 – 08/2009	Research Group Leader (Systems Biology and Mathematical Modelling), Max Planck Institute of Molecular Plant Physiology, Potsdam
01/2007 – 03/2007	Research Fellow (Computational Systems Biochemistry), Charite Berlin
04/2003 – 12/2006	Postdoctoral Research Fellow (Theoretical Biophysics), Humboldt University Berlin
08/1998 – 04/2003	PhD in Theoretical Biophyics
09/1996 – 06/1998	IT Consultant, Control Data GmbH, Frankfurt(Main)
04/1996	Graduated in Mathematics and Physics

Lizbeth Hedstrom

Current Position: Markey Professor of Biology and Chemistry

Education and Training

1989–91	University of California, San Francisco
1986–89	University of California, San Francisco
1985–86	Massachusetts Institute of Technology

1980 – 85   Brandeis University, Ph.D. Biochemistry 1986
1976 – 80   University of Virginia, B.S. Chemistry 1980 with high distinction

Honors

Searle Scholar (1993); Beckman Young Investigator (1995); NSF CAREER Award (1995); Louis Dembitz Brandeis Prize for Excellence in Teaching (2007); AAAS Fellow (2010)

Research Interests

Structure/function relationships in enzyme catalysis. Serine proteases. Purine metabolism. The design of enzyme inhibitors as mechanistic probes and potential pharmaceutical agents. Pathological mechanisms of retinitis pigmentosa. Protein degradation

Martin G. Hicks

is a member of the board of management of the Beilstein-Institut. He received an honours degree in chemistry from Keele University in 1979. There, he also obtained his PhD in 1983 studying synthetic and theoretical approaches to the photochemistry of pyridotropones under the supervision of Gurnos Jones. He then went to the University of Wuppertal as a post-doctoral fellow, where he carried out research with Walter Thiel on semi-empirical quantum chemical methods. In 1985, Martin joined the computer department of the Beilstein-Institut where he worked on the Beilstein Database project. His subsequent activities involved the development of cheminformatics tools and products in the areas of substructure searching and reaction databases.

Thereafter, he took on various roles for the Beilstein-Institut, including managing director-ships of subsidiary companies and was head of the funding department 2000 – 2007. He joined the board of management in 2002 and his current interests and responsibilities range from organization of Beilstein Symposia with the aim of furthering interdisciplinary com-munication between chemistry and neighbouring scientific areas, to the publishing of Beil-stein Open Access journals such as the Beilstein Journal of Organic Chemistry and the Beilstein Journal of Nanotechnology.

Carsten Kettner

studied biology at the University of Bonn and obtained his diploma at the University of Göttingen. In 1996 he joined the group of Dr. Adam Bertl at the University of Karlsruhe and successfully narrowed the gap between the biochemical and genetic properties, and the biophysical comprehension of the vacuolar proton-translocating ATP-hydrolase using the patch clamp techniques. He was awarded his Ph.D for this work in 1999. As a post-doctoral student he continued both the studies on the biophysical properties of the pump and investigated the kinetics and regulation of the dominant plasma membrane potassium channel (TOK1).

In 2000 he moved to the Beilstein-Institut to represent the biological section of the funding department. Here, he is responsible (a) for the organization of the Beilstein symposia (ESCEC, Glyco-Bioinformatics and "the" Beilstein Symposium) and the publication of the proceedings of the symposia and (b) for the administration and project management of funded research projects such as the Beilstein Endowed Chairs (since 2002), the collaborative research centre NanoBiC (since 2009) and the Beilstein Scholarship program (since 2011). In 2007 he started a correspondence course at the Studiengemeinschaft Darmstadt (a certified service provider) where he was awarded his certificate of competence as project manager for his studies and thesis. Since 2004 he coordinates the work of the STRENDA commission and promotes along with the commissioners the proposed standards of reporting enzyme data (www.strenda.org). These reporting standards have been adopted by 28 biochemical journals for their instructions for authors and are subject for the development of an electronic data capturing tool. Against this background, in 2011, Carsten was appointed to co-ordinate another standardization project (MIRAGE) which is concerned with the uniform reporting and representation of glycochemistry data in publications and databases.

Thomas Millat

1993 – 1998	Study of Physics, University of Rostock
1999	Dipl. Phys., University of Rostock
1999 – 2003	Research associate, Chair in Quantum Statistics of Many Particle Systems, University of Rostock
2003	Dr. rer. nat. in Theoretical Physics
2003 –	PostDoc, Systems Biology and Bioinformatics Group, University of Rostock
2007 –	Principal Investigator for European transnational SysMo projects COSMIC and Bacell

Research Interests

Dynamic adaptation and stress response of microorganisms;
Approximations and assumptions in the dynamic modeling of biological systems;
Measures and characteristics of dynamic signals and their application in the analysis of cellular systems;
Design principles of biochemical networks

Frank M. Raushel

received a B.A. degree in chemistry from the College of St. Thomas in Minnesota, and a Ph.D. degree in biochemistry from the University of Wisconsin in 1976. After postdoctoral training in biophysics at Penn State University, he joined the faculty at Texas A&M University in 1980. He is currently Distinguished Professor of Chemistry and Davidson Professor of Science. The research in his laboratory focuses on the elucidation of enzyme reaction mechanisms, evolution of enzyme active sites, and the discovery of new enzyme catalyzed reactions. The Raushel group initiated the use of a bacterial phosphotriesterase as a model system for the activation of water by binuclear metal centers and as a template for the rational redesign of an enzyme active site for the stereoselective hydrolysis of chiral organophosphate nerve agents. His laboratory is currently working to develop effective strategies for elucidating the catalytic properties of enzymes of unknown function. He is the recipient of the 2009 Repligen Award from the American Chemical Society.

Vladimir Sobolev

2004 –   Senior Staff Scientist, Department of Plant Sciences, Weizmann Institute of Science (WIS), Rehovot, Israel

Education

1976   Ph.D., Physical Chemistry, Institute of Catalysis, Siberian Branch of the Acad. of Sci. of USSR.
1970   M.Sc., Physics, Department of Physics, Novosibirsk State University, USSR.

Professional Experience

1991 – 04   Scientist, Giladi Fellow, Associate Staff Scientist, Department of Plant Sciences, WIS, Rehovot, Israel
1984 – 91   Scientist, Laboratory of Modeling of Intermolecular Interaction in Biological Systems, Institute of Experimental and Theoretical Biohpysics, Pushchino, USSR.
1970 – 84   Scientist, Laboratory of Physics of Enzyme Systems, Institute of Biophysics, Pushchino, USSR.

Scientific Productivity

1 book, 63 publications (papers in refereed journals, book chapters, reviews), more than 50 invited talks, author (PI) of bioinformatics tools for structure analysis and prediction (http://ligin.weizmann.ac.il/space)

Reinhard Sterner

Education

1988	Diploma in Biology, University of Munich
1991	Dissertation in Biology, University of Munich
1996	Habilitation in Biochemistry, University of Basel, Switzerland

Career

1988 – 1991	Scientific collaborator, Institute of Zoology, University of Munich
1991 – 1997	Postdoctoral associate, Department of Biophysical Chemistry, Biozentrum, University of Basel, Switzerland
1997 – 1999	Heisenberg fellow, Institute of Microbiology and Genetics, University of Göttingen
1999 – 2003	Professor (C 3) of Biochemistry, Department of Chemistry, University of Cologne
2004 – present	Professor (C 4) of Biochemistry, University of Regensburg

Awards, Grants

| 1998 – 1991 | PhD fellowship (state of Bavaria) |
| 1997 – 1999 | Heisenberg fellowship (DFG) |

Committees

Since 2009	Spokesman of the GBM interest group "Protein engineering and design"
Since 2006	German representative of the "International Network of Protein Engineering Centres"
Since 2004	Coordinator of the DFG priority program 1170 "Directed evolution to optimize and understand molecular biocatalysts"

Research Interests

Protein design; structure, function and evolution of enzymes; functional annotation of new enzymes

Christian P. Whitman

Education

| 1979 | B.S. (Chemistry), University of Connecticut, Storrs, |
| 1984 | Ph.D. (Pharmaceutical Chemistry), University of California, San Francisco, 1984 |

Positions and Employment

1987 – 1993	Assistant Professor, Medicinal Chemistry, University of Texas at Austin
1993 – 1998	Associate Professor, Medicinal Chemistry, University of Texas at Austin
1998-present	Professor and Division Head, Medicinal Chemistry, University of Texas at Austin
2001 – 2007	Graduate Adviser, College of Pharmacy, University of Texas at Austin
2007-present	Graduate Adviser, Division of Medicinal Chemistry, University of Texas at Austin
2001 – 2007	Graduate Advisor, College of Pharmacy, The University of Texas at Austin
2001 – 2005	Jacques P. Servier Regents Professorship in Pharmacy
2005-present	Romeo T. Bachand, Jr. Regents Professorship in Pharmacy

Other Experience and Professional Memberships

2000	Co-Chair, GRC on Enzymes, Coenzymes, and Metabolic Pathways
2001 – 05	Member, NIH Biomedical Research Training (BRT) Study Section
2002-present	Editor, Bioorganic Chemistry
2006 – 12	Chair, Texas Enzyme Mechanisms Conference, Austin, TX
2008 – 12	Member, NIH Molecular Structure and Function E Study Section
2010 – 12	Chair, NIH Molecular Structure and Function E Study Section

Honors

1985 – 1987	NIH Postdoctoral Fellowship (Biochemistry)
1993 – 1994	Texas Excellence Teaching Award
2000 – 2001	Texas Excellence Teaching Award
2006	Fellow of the American Association for the Advancement of Science

Ulrike Wittig

studied biochemistry at the University of Leipzig, Germany, and received a Ph.D. in biology from the University of Heidelberg, Germany, in 1998 for the experimental work on mechanisms of apoptosis and oxidative stress in mammalian cells. In 1998, she joined the EML Research gGmbH in Heidelberg, Gemany as a research associate in the Scientific Database and Visualisation group. In 2010 EML Research gGmbH was restructured and she now works at the new established Heidelberg Institute for Theoretical Studies (HITS). Here she is mainly involved in the development of biological databases, database curation and annotation. She works as a database curator of the SABIO-RK database for biochemical reactions and their kinetics (http://sabiork.h-its.org) developed at HITS.

Her research interests include data curation, integration and standardisation in biological databases and information extraction from biological data sources.

169

Authors' Index   Experimental Standard Conditions of Enzyme Characterization, September 12th – 16th, 2011, Rüdesheim/Rhein, Germany

Index of Authors

Index

171

Index   Experimental Standard Conditions of Enzyme Characterization, September 12th – 16th, 2011, Rüdesheim/Rhein, Germany